THE SACRED GEOMETRY OF
ANCIENT GREECE

THE SACRED GEOMETRY OF ANCIENT GREECE

by

by Lucie Wolfer-Sulzer

Translated by Bernhard Rest

eglantyne books

Published by Eglantyne Books ltd,
The Club Room, Conway Hall, 25 Red Lion Square, London WC1R 4RL.
www.eglantynebooks.com
Translation and notes Copyright © 2024 by Eglantyne Books Limited.
All rights reserved
ISBN 978-1-913378-16-5
Printed in the UK by Imprint Digital ltd.
A CIP record for this title is available from the British Library.
Book layout and design by Eric Wright
Production team:
Robert and Olivia Temple, Michael Lee and Eric Wright

CONTENTS

THE AUTHOR

The Swiss scholar Lucie Wolfer-Sulzer was born Lucie Sulzer on July 23, 1885, and died July 25, 1946. She married Dr. jur. Leo Heinrich Wolfer (1882-1969), and thereafter was known as Lucie Wolfer-Sulzer. Her reverence for Greek antiquity began in her childhood under the influence of her famous grandfather, Dr. Friedrich Imhoof-Blumer (1838-1920), a historical numismatist who was the world's leading authority on ancient Greek and Roman coins. Countless priceless images of lost buildings, portraits of historical personalities, and mythological motifs, were discovered and published by him as the images preserved on coins. Lucie also became perfectly fluent in ancient Greek. She first became a historian of arts and crafts who published, with Gret Hasler, a detailed study of traditional Swiss lace patterns, *Kreuzstich und Filetmuster aus Graubünden* (*Cross Stitch and Fillet Patterns from the Canton of Graubünden*, Chur, Switzerland, 1929), illustrated with 80 photographs. It was inevitable that her work in collecting and studying the lace patterns turned her attention towards the geometry of design. She also was interested in the ancient history of musical theory, concerning which she published two booklets: *Pythagoreische Töne* (*Pythagorean Tones*, 1943), which relates to sacred geometry and appears as Part Three of this book, and *Das Griechische Tonsystem* (*The Greek Musical System*, 1946), which was published in the year she died and appears as Part Four of this book. The two booklets were first drafts which she privately printed and circulated to colleagues, to elicit their comments. Unfortunately, she died in the same year as the second booklet appeared, and was unable to produce a finished work on the subject as she had intended.

As part of her love of Greek antiquity, Lucie took an interest in how the Greeks designed their sacred temples and made their statues so lifelike. Under the influence of the Romanian polymath Matila Ghyka, she came to realize the importance of the Golden Section in these Greek achievements. And when she discovered the works of the American Jay Hambidge, such as his book about the Parthenon, she realized that the other missing piece of the puzzle had been found: *dynamic symmetry*, a technical and precise geometrical approach to design which Hambidge, as a practising artist, had rediscovered on his own. Later it came to light that the name which Hambidge had chosen for it was exactly the same name by which it had been known in antiquity. Lucie published her first book about this in 1939: *Das Geometrische Prizip der Griechisch-Dorischen Tempel* (*The Geometrical Principle of the Greek-Doric Temples*). It was a result of her trip through Sicily, where she studied and measured and drew the plans of some of the Greek temples. She followed this book in 1941 with *Urbild und Abbild* (*Archetype and Image*), which treats of the application of Greek sacred geometry to statues. It is these two books, as well as the two booklets on musical theory, which are published together here, all translated into English for the first time. Part One is *Archetype and Image*, which although it was published second, is somewhat easier of access to the reader, and hence a better beginning to the understanding of Lucie's work. Her studies of the Greek temples thus appears as Part Two.

There appear to be no surviving Wolfer-Sulzers, and it has not been possible to obtain a photo of the author or any further information about her.

The author is deeply grateful to her grandfather,

DR. FRIEDRICH IMHOOF-BLUMER

(Numismatist, 1838 – 1920),

for instilling in her a reverence for Greek culture at a very early age. The author has arrived at the results outlined in these pages as a layperson and without any prior training.

Winterthur, March 1941

PART I

ARCHETYPE AND IMAGE
OF THE ANCIENT GREEK FORM

"The lawmaker must leave no stone unturned…
to reinforce the notion that law and
art themselves are derived from nature."
Plato: *Laws*, Book X, 890d

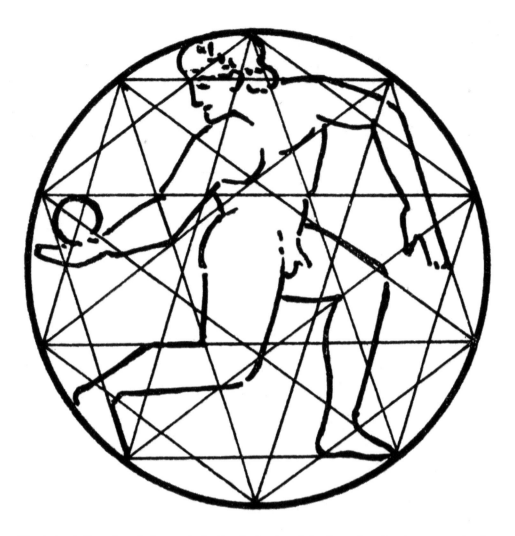

The design is that of a painting on the inside of a bowl made by the ancient Greek potter Kachrylion.
From Edmond Pottier, *Vases Antiques du Louvre.*

I.

"Pythagoras's teachings are based on measurements and numbers." Every schoolbook contains this sentence. [*This was the case at the time of writing.*]

In our modern age it can nevertheless be difficult to fathom why mathematics enjoyed such high priority in ancient Greece and the extent to which it permeated philosophy and art. The Pythagoreans, much like the Orphics and Gnostics, kept the essence of their worldview – the divine unnameable – strictly secret. Apart from the kaleidoscopic, allegoric realm of the gods, it was merely numbers, geometric figures and proportions (and by no means all of those) that were permitted to reach the general public. Only Plato's writings began to provide clues about Pythagorean knowledge. His work *Timaeus* contained a description of the world as a whole and some mathematical information, most of which, however, remained fragmented and mysterious, which gave rise to endless speculation.

The present study will cite Plato in a confirmatory rather than a directive manner. Matila Ghyka, the author of valuable works on geometry[1][2], has the following to say about him:

"On pourrait être tenté de croire que Platon non seulement avait prêté le serment du secret, mais l'avait bien gardé, ne laissant échapper que des étincelles de la grande lumière."[3] [Ghyka: *We may after all be tempted to believe that Plato did indeed take a vow of secrecy, and kept it well, only letting a few glimmers of the great light shine through to mark out the pathways for those who, through the ages, would be worthy to transmit the torch.*] In the same chapter, one can find this important passage: "Ces serments (du silence et du secret) étaient du reste usuels non seulement dans les sects d'initiés affiliés aux mystères, mais même dans les confréries corporatives ou professionelles."[4] [Ghyka: *These oaths (of silence and secrecy) were common, both in the cults of initiates affiliated to the mystery cults (Eleusinian, Orphic, etc.) and in the corporative or professional brotherhoods ...*]

The fact that artists and craftsmen were adherents of secret doctrines explains why – as Jacob Burckhardt mentions – Greek sculptors and architects are of curiously low prominence compared to statesmen and philosophers. And so the formal laws of sculpture and architecture remained shrouded in the great secret. Only the legend of the [lost] *Canon of Polykleitos* has survived, as well

---

1 *Esthétique des Proportions dans la nature et dans les arts* [*The Aesthetics of Proportion in Nature and in the Arts*], Gallimard, Paris, 1927.

2 *Le Nombre d'Or* [*The Golden Number*], 2 vols., Editions de la nouvelle revue française, Gallimard, Paris, 1931.

3 As evidence, Ghyka refers to Plato's *Seventh Letter*. [*Le Nombre d'Or*, Vol. II, p. 21.]

4 *Le Nombre d'Or.* Chapter VI, Pythagore, p. 24. [*Note :* This references is erroneous. The correct reference is to the first Chapter of Volume Two, p. 24. By a printing error in the original, this chapter (really the seventh of the whole work) is headed 'Chapitre VI' instead of 'Chapitre I', whereas the chapter following it is headed Chapitre II.]

as a quotation from Vitruvius, in which he claims that the Greek temples were designed according to the proportions of the human body.

It is no coincidence that a lack of a profound classical foundation for artistic construction was perceived especially during the Renaissance era. Even Leonardo da Vinci and Dürer pursued this conundrum, hoping that its solution would provide ultimate consummation.

A legend exists, according to which Hippasus [of Metapontum, c. 530 – c. 450 BC], one of Pythagoras's pupils, betrayed the secret of the pentagram and dodecahedron. [This was claimed by Iamblichus, the neo-platonic philosopher.] As his punishment, the gods made him perish at the bottom of the sea. Thus it was the pentagram and the dodecahedron that were associated with the divine. The five-pointed star, the pentacle, was surrounded by an aura of magic until the late Middle Ages.

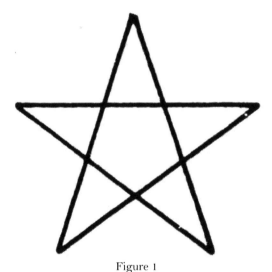

Figure 1

The pentagram was considered to be the signum of the Pythagoreans, and this geometric figure is, in turn, the symbol of irrational proportions within the absolute order. Even though its properties must have been known before Pythagoras, it was he – according to the testimony of Eudemus [of Rhodes, c. 370 – c. 300 BC, a pupil of Aristotle] – who discovered the irrational.

The irrational – incommensurability – exists in metric proportions that cannot be expressed in numbers.

Examples:

1. The ratio between the side of a square and its diagonal: 1 : 1.4142 or 1 : $\sqrt{2}$

Figure 2. The square and its diagonal.

2. The ratio between the side of a square and the diagonal through two squares: 1 : 2.236 or 1 : $\sqrt{5}$

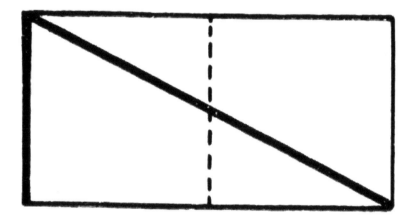

3. The Golden Ratio or Divine Proportion:

$$0.618 : 1 = 1 : 1.618 = 1.618 : 2.618$$

or $a : b = b : (a+b) = (a+b) : (a+2b)$

The proportion $1 : \sqrt{5}$ and the Golden Ratio are closely interconnected: $\sqrt{5} = 2.236 = 0.618 + 1.618$

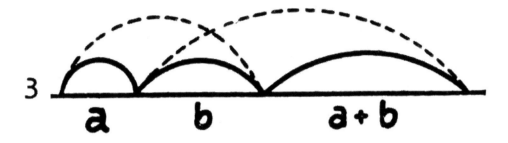

Figure 3.

The concept of decimal fractions was unknown in Antiquity. It is generally assumed that for this reason immeasurability was perceived to be part of the "divinely unnameable" and was therefore subject to the laws of secrecy.

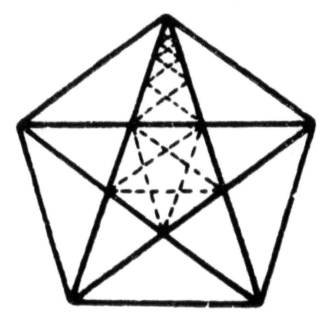

Figure 4.

Taken together, the pentagon and the pentagram [Fig. 4] produce the principle of perfect proportional increase and reduction. Not only do they wane infinitely towards their shared centre and grow outwards in an ever-increasing circle, but, lying at an angle of 36° to each other, the five points of the pentagram themselves replicate the manifestations of the basic forms, namely the pentagon, the rhombus and the triangle, decreasing or increasing in size according to the direction of their position. The diagonals of a pentagon intersect each other (and the radius of the circumscribing circle) at the "Golden Ratio".

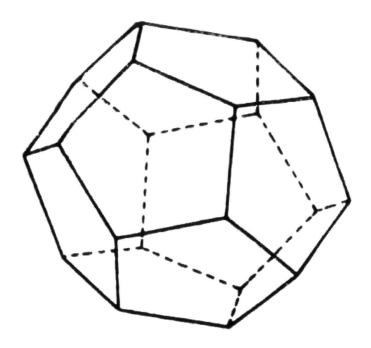

Figure 5. The dodecahedron.

The surfaces of twelve equilateral pentagons combine to form a regular solid figure, the dodecahedron [Figure 5]. If one connects its vertices diagonally with straight lines, twelve pentagrams are created on both the inside and the outside of the body. The external pentagrams can be grouped into five intersecting cubes that are contained within the solid body, while the other, internal diagonals through their intersections and segmentations form an icosahedron, whose twenty faces are equilateral triangles. Through the extension of the edges of the dodecahedron in an outward direction one can construct a SMALL STELLATED DODECAHEDRON [Figure 6]. In two dimensions, that is to say, projected onto a plane, the small stellated dodecahedron appears as a DOUBLE PENTAGRAM, i.e. as two overlapping pentagrams that are inscribed in a circle whose circumference is divided into ten parts, namely the decagram.

Figure 6. The small stellated dodecahedron, which when portrayed on a flat surface appears as two
overlapping pentagons, one pointing up and one pointing down.
(Additionally, a small pentagon pointing down is seen in the centre.)

For the sake of completeness, the four other regular geometric bodies – and their important symbolic meaning in Antiquity, representing the four elements – should be mentioned here. Plato writes: *"To earth, let us assign the cubic form… Furthermore, we assign the smallest body, the tetrahedron, to fire, the greatest, the icosahedron, to water, and the intermediate, the octahedron, to air."* In comparison, it is striking how very briefly he touches on the last one of the regular geometric bodies, the dodecahedron: *"And seeing that there still remained one other compound figure, the fifth, GOD USED IT UP FOR THE UNIVERSE IN HIS DECORATION THEREOF."*

ἔτι δὲ οὔσης συστάσεως μιᾶς πέμπτης, ἐπὶ τὸ πᾶν ὁ θεὸς αὐτῇ κατεχρήσατο ἐκεῖνο

διαζωγραφῶν. *Timaeus,* 55c

["There was a fifth combination which God used in the delineation of the Universe.'"– Benjamin Jowett translation; this translation completely obscures Plato's meaning by speaking of a "combination" without explaining that it was a figure, i.e. a "Platonic solid", namely the dodacahedron.

"And seeing that there still remained one other compound figure, the fifth [the dodacahedron], God used it up for the Universe in his decoration thereof." – R. G. Bury translation.

'There was yet a fifth combination, (the regular dodecahedron); and this the Deity employed in tracing the plan of the universe.' – Henry Davis translation for Bohns Library, p. 363; "plan" is better than "decoration"!]

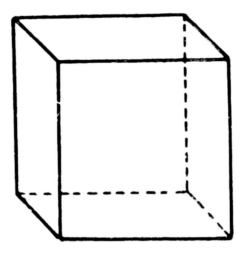

The cube according to Plato "corresponds to earth".

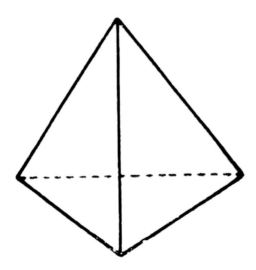

The tetrahedron according to Plato "corresponds to fire".

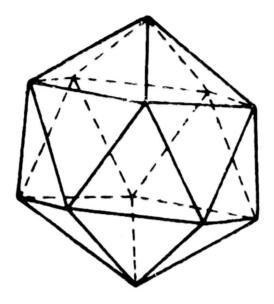

The icosahedron according to Plato "corresponds to water".

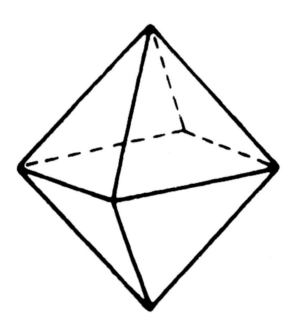

The octahedron according to Plato "corresponds to air".

PART I: Archetype and Image

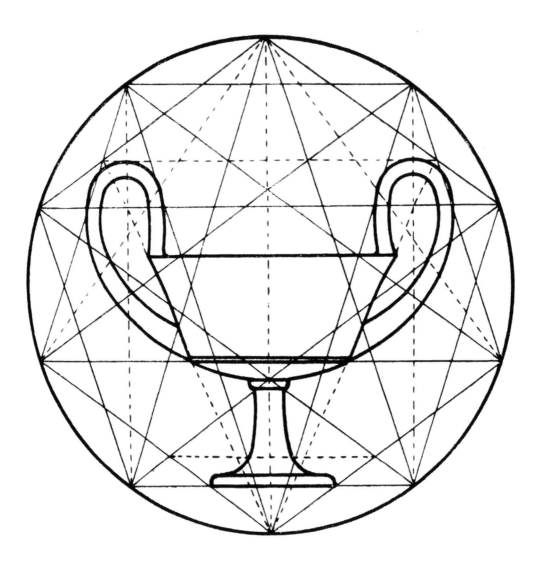

Kantharos in the Boston Museum, from Jay Hambidge, *The Greek Vase.*

II.

Since time immemorial, architecture has been based on fundamental forms; artistic constructions were always first represented as sketches. Matila Ghyka wrote these apt words about the transfer of a proportioned form to the sketch:

"L'essai de réaliser une pulsation de croissance à trois dimensions, à trois temps, une proportion théorique idéale pour la croissance des volumes, n'aboutit pas à un résultat pratique, car on s'y butte, comme pour le problème de la duplication du cube, à une équation du 3e degré irrésoluble euclidiennement. C'est une indication de plus pour se contenter de graphiques plans et de commodulations de surfaces pour le traitement pratique des questions de proportions concernant les volumes." [Le Nombre d'Or, Matila Ghyka, Volume One, Chapter Four, p. 86, remarks found within Footnote 1, which refer back to Ghyka's earlier work, Esthétique des Proportions dans la Nature et dans les Arts (1927), Chapters 5 and 7.] [Ghyka: "I show that the attempt to realise a three-dimensional, three-beat pulsation of growth, an ideal theoretical proportion for the increase or growth of volumes, cannot lead to a usable result, because one will come up against a third-degree equation ($x^3 = x^2 + x + 1$) – as in the problem of the duplication of the cube – that is impossible to resolve using 'Euclidean' methods. That is one more reason why we should content ourselves with graphical plans and 'commodulation' of surfaces for the practical treatment of questions of proportions applied to volumes."]

The quote is from Matila Ghyka, *Le Nombre d'Or*, p. 86.

In the early Christian and the medieval periods, it was the two most basic geometric forms, the triangle and the square, that generally served as the basis for the greatest architectural edifices. The ground plans for basilicas and Byzantine churches were developed from the square and octagon, probably because the symbol of the cross can be expressed most clearly using these forms, while during the Gothic period the equilateral triangle and hexagon were preferred. In either case, the DOUBLING of the basic shape within a circle plays a crucial role in the refinement of the gradation of the form.

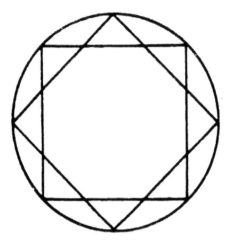

The doubling of a square within a circle.

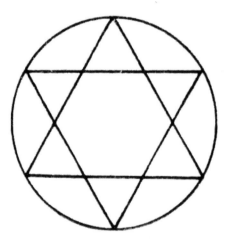

The doubling of a triangle within a circle (one pointing up and one pointing down).

The decagram.

The doubling of the pentagon, on the other hand, was all but forgotten for two millennia. Only the more humble decagram – a part of the complete fundamental form – found some use now and then.

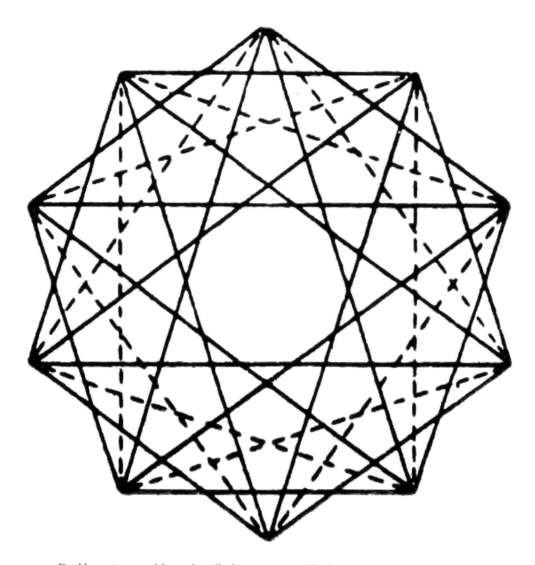

Double pentagons with two inscribed pentagrams and a decagram (shown by dotted lines).

The fact that the basic pentagram does not provide a satisfactory basis is demonstrated by the countless and repeated attempts at its interpretation and application. The Pythagoreans' emblem reveals its deep dual meaning through the surprising results obtained when one completes the basic form. It is remarkable, how DUALITY is generally emphasised in Greek culture. One example, which is directly connected to the questions posed here, is the well-known oracle of Apollo and its demand: "Double the cube-shaped altar of your god!" This does not mean a simple addition. In addition, as Ghyka mentions, the doubling of a cube does not lead to a uniform development, as is also the case with the dodecahedron. However – much like a doubled pentagram as the two-dimensional representation of a small stellated dodecahedron – a double cube, projected onto a plane and inscribed into a sphere as an octagon, possesses the inherent potentiality of increment and decrement (in the ratio of 1 : √2). Thus it leads to the indivisible, the irrational – to the divine.

From the 6th century BC onwards, this irrational property of metric ratios within regular geometric figures determined proportions in Greek architecture. [*Editor's note: it was at this time that Greeks were permitted to enter Egypt, and that was the source of this technique amongst the Greeks.*] The ratio of 1 : √2 is only found in exceptional cases in larger buildings. One such example is the Telesterion at Eleusis, where annual mystery cult rituals were held. The vast majority of temples – from the early Doric examples to the Hellenistic buildings of Asia Minor – are based on the ratio of 1 : √5, thereby basing their measurements and proportions on the double pentagram. Yet even this principle turns out to be a complex affair. The fundamental figure remains hidden; its structure is transformed into a system of rectangular planes that is more suitable for architectural purposes.

During the first decades of our century, the American artist and mathematician Jay Hambidge newly discovered this TRANSLATION of the Greek basic form and its application for composition and analysis. [Books by Jay Hambidge (1867-1924) include *The Diagonal (published in a series of 12 parts, 1919-1920), Dynamic Symmetry: The Greek Vase* (1920), *The Parthenon and Other Greek Temples: Their Dynamic Symmetry* (1924), *The Elements of Dynamic Symmetry* (1926, posthumous), *Dynamic Symmetry in Composition as Used by Artists* (1923), and *Practical Applications of Dynamic Symmetry* (1932, posthumous) , all published by Yale University Press, USA. See also Christine Herter, *Dynamic Symmetry: A Primer*, Norton, New York, 1966.]

The present study could only be developed thanks to the clear and exact results achieved using Hambidge's method (referred to by him as "Dynamic Symmetry"). In actual fact, during the course of this investigation we will slowly feel our way backwards, from reference point to reference point, from Greek architecture to the archetype of Pythagorean thought.

"Dynamic Symmetry" emphasises proportional expansion. It is based on the proportion between lengths and the irrational values of √2, √3 and √5 and unfolds as an areal composition of square and rectangle and their diagonals. Hambidge's attempt to explain harmony within classical Greek architecture by using his segmentations is plausible but not exhaustive; his death in 1926 sadly forestalled the completion of his studies.

The structural proportions of even the oldest Doric temples in Sicily can be clearly categorised using Hambidge's method. Starting with the main dimensions (length and width in the horizontal section, and width and height in the vertical section), the plans can be subdivided into geometric values of similar proportions on the basis of which the edifices are designed. Further consistent analysis leads down to the smallest details, such as the profiles of cornices or the structure of other ornaments, seamlessly integrating these into the whole. As Hambidge always gives the aspect ratios of his elemental and compound rectangles as decimal fractions, a simple conversion is all that is required to prove his method. A certain inconsistency, the full investigation of which would exceed the scope of this investigation, deserves a brief mention. Although – strictly scientifically speaking – they date from a pre-Pythagorean era, Archaic buildings and statues already display features associated with the Pythagorean School.

[The four drawings opposite exemplify a technique used by the Greeks after the 6th century BC and rediscovered by Hambidge. He (and they) took the diagonals of, first a square, and then a succession of rectangles and swung them down to the horizontal, thus creating a series of extended rectangles whose larger sides had the irrational numerical values of $\sqrt{2}$, $\sqrt{3}$, $\sqrt{4}$, and $\sqrt{5}$. Hambidge called these successively by the names of "the root 2 rectangle", "the root three rectangle", "the root four rectangle", and "the root five rectangle". For a full understanding of the truly astonishing discoveries of Jay Hambidge it is necessary to consult his own books, some of which were published posthumously by his widow. The "root two rectangle" is created from a square; the diagonal of the square creates the longer side of the rectangle, and this process is then extended to a succession of further rectangles as shown in the drawings.]

The following diagrams show the underlying geometry of the ground-plans of the temples.

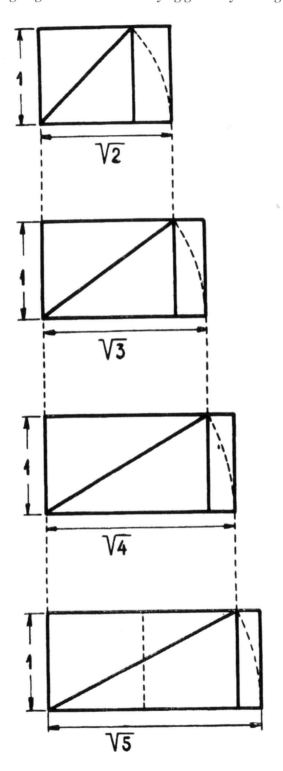

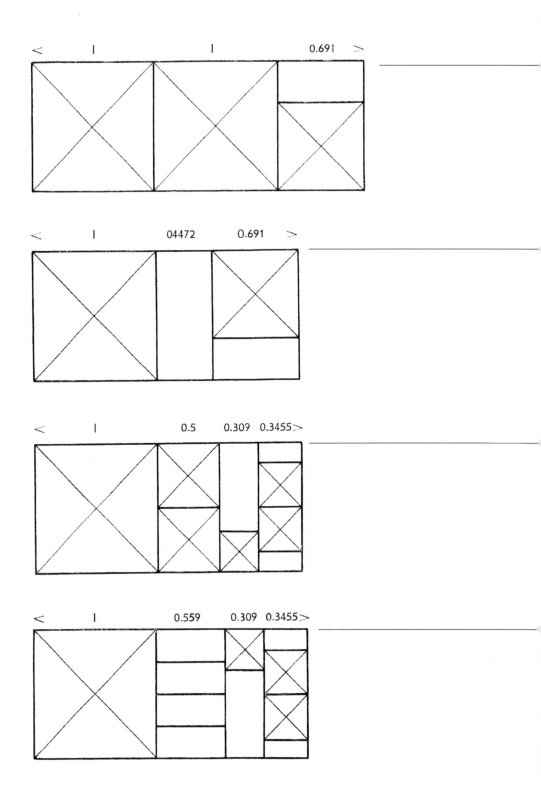

The ground plans of four temples adhere to the following schema:

Temple C, Selinunt
Width   71.12 metres
Length   26.43 metres

Concordia Temple, Agrigento
Width   19.75 metres
Length   42.23 metres

So-called Basilica, Paestum
Width   26.00 metres
Length   55.76 metres

Zeus Temple, Olympia
Width   30.30 metres
Length   66.74 metres

The following attributes should be highlighted:

1.   The constant and exclusive use of the two basic forms, square and rectangle $1 : \sqrt{5}$

2.   The emphasis on the double square and of the compound form square + rectangle $1 : \sqrt{5}$

3.   A primary square whose side length is equal to the total width of the structure

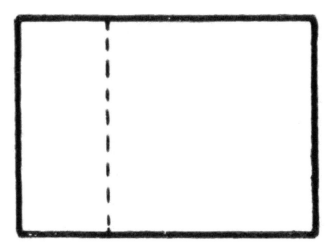

The primary square at right extended as a rectangle to the left.

Hambidge did not address the question of the significance of this striking arrangement in the geometric plans. However, such rigorous segmentation requires an explanation of the fundamental creative idea.

The initial motive of the present study was to determine the meaning of the primary square. A simple segmentation of this fundamental square for the vertical elevation is trivial and in no way exhaustive. However, when a circle and double pentagram are inscribed, forming a UNION, it reveals itself as a KEY FIGURE, which serves to predetermine all measurements that are derived from the combinations of a square and a $\sqrt{5}$ rectangle [1].

In practice, the key figure not only confirms a construction based on rectangular geometry, but, as a principle and allegory, it also leads from duality [2] to the triad, to a kind of trinity [3].

A temple's plan can therefore always be represented in two ways: either within a circle and double pentagram, or as developed from square + rectangle.

The small Temple of Aphaia on the island of Aegina is one of the most beautiful examples of classical Greek architecture.

The temple of Aphaia on the Island of Aegina.

This temple's rectangular arrangement is derived from the following schema (its elevation is based on the proportions of the primary square):

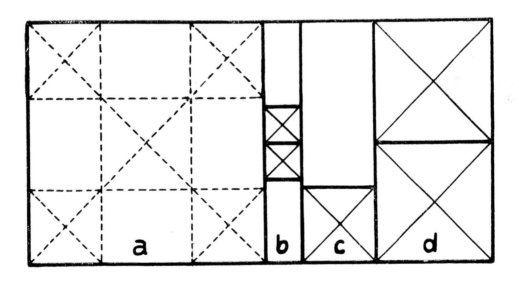

Ground plan:

Length: 30.50 metres

Width: 15.53 metres [Within the primary square, the rectangles overlap 1 : $\sqrt{5}$.]

The numerical values of the proportions of the long side:

$$a=1, \; b=0.1545 \; \left[\frac{1}{2\,(\sqrt{5}+1)}\right], \; c=0.309 \; \left[\frac{1}{\sqrt{5}+1}\right], \; d=0.5 \; \left[\frac{1}{2}\right].$$

Arithmetical verification: ratio of the ground plan 1 : 1.9635

15.53m x 1.9635 = 30.493m

[1]   By drawing horizontal and vertical lines through the points and intersections of the double pentagram (which is inscribed into the circle and square), rectangular areas are created that consist of squares + rectangles 1 : $\sqrt{5}$.

[2]   Square + rectangle 1 : $\sqrt{5}$

[3]   The unity of circle, square and rectangle 1 : $\sqrt{5}$ (see the second Figure in the series above), with a circle and double pentagram inscribed into the square. This rectangle (with a side ratio of 1 : 1.4472) is the geometric form that most often occurs in the combinations of the temple's dimensions.]

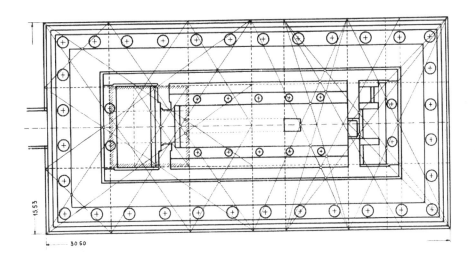

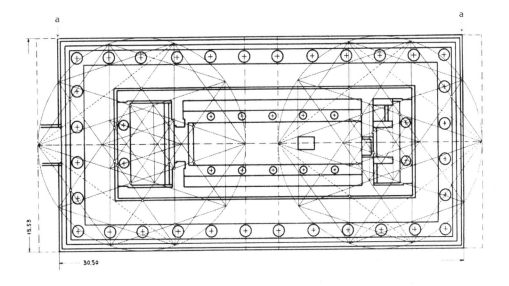

Ground plan

[4] Built circa 480 BC

Development of the segmentation according the schema described above. Segmentation using the double pentagram. The Temple of Aphaia on the island of Aegina. On the plan, the double pentagram corresponds better to the actual proportions of the building when the side length of the pentagon is taken as the starting point. The reasons for this approach will become clear later.

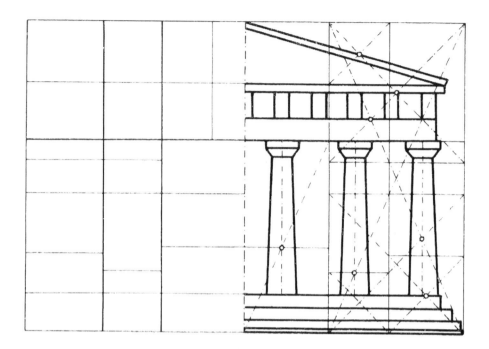

Elevation

The Temple of Aphaia on the island of Aegina

Width of the structure: 15.53 metres

Height of the structure: 10.73 metres

Arithmetical verification

Ratio of the elevation:

$$1:0.691 \left[\frac{\sqrt{5}}{\sqrt{5}+1} \right]$$

15.53m x 0.691 = 10.731m

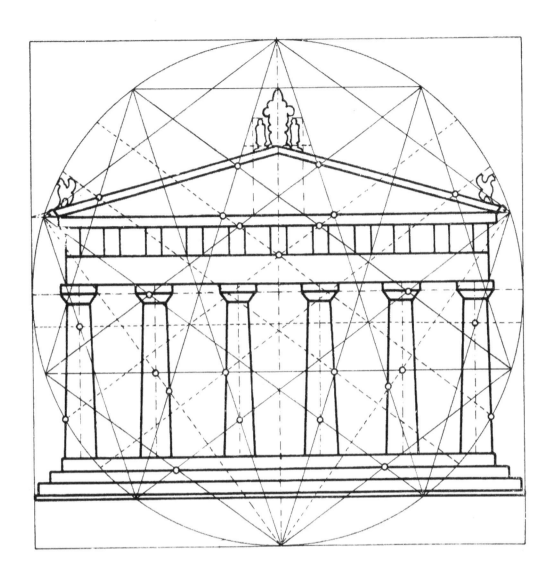

When using the 10-pointed star as a geometric basis, its segmentation must be executed as clearly as possible. In particular, it is vital to distinguish between the outer and inner double pentagons (the "major" and the "minor") and the double pentagram inscribed within them.

Since the metric proportions of the building are largely based on intersection points, certain axes and lines within the decagram as well as the continuation of the diagonal of the "minor" to the circular periphery are required for the completion of the fundamental framework.

[We turn now to an example from the Acropolis at Athens.]

The Caryatid Porch

The following results are derived from an analysis of the Caryatid porch of the Ionic Erechtheion on the Acropolis of Athens (using a photograph by the English archaeologist [Francis] Penrose).

The image of the elevation within the circle and decagram is simple and clear. The female figures, the caryatids, align remarkably well with the inner pentagram. A detailed examination of a single statue (within a correspondingly reduced fundamental form) clearly reveals its connection to the whole. [See Plate I in the Appendix]

Thus Vitruvius's fabled lore becomes fact: THE CARYATIDS ON THE ERECHTHEION CONFIRM THE TRANSFER OF HUMAN PROPORTIONS TO ARCHITECTURE.

From here it is now possible to restore the Canon of Polykleitos, the famous ancient sculptor. However, in order to compile a conclusive survey – an idealised classification of human proportions – it will be necessary to carry out further measurements of original sculptures.

At first, we shall examine upright, unclothed figures, which reveal the bodily proportions most readily.

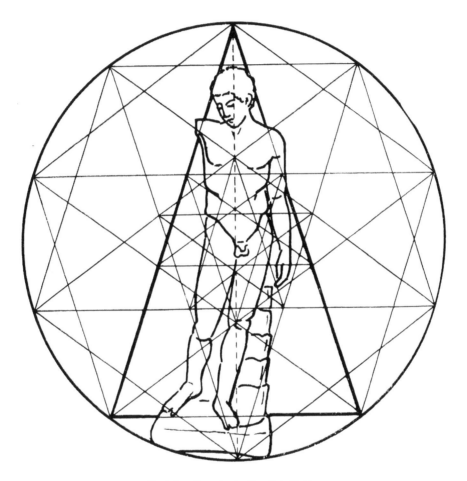

Kyniskos (London), after Polykleitos

In the double pentagram, they stand between the horizontal sides of the two pentagons, that is to say, they do not meet the surrounding circle. The correspondence to the spatial arrangement of the temples is self-evident.

An equilateral triangle of axes, which is formed by the uppermost point of the decagram and the opposing side of the pentagon (referred to as the *"triangle sublîme"* in Ghyka's writings), is all that is required for the integration of the human form. In addition to the lateral proportions, the angles of 36°, 72° and 108° now gain great importance. Decorative adjuncts to the statues, such as Doryphoros' staff [Plate II in the Appendix], the mace of Heracles (in London), vases, buttresses, animals, etc., conform to the overall composition and affirm the fundamental form. The decagram was determinative also for pre-Polykleitan statues, but these are still of a rather clunky composition and Egyptian austerity. An Archaic statue by Polymedes of Argos (Delphi), which bears an enigmatic engraving on its torso, deserves special attention [Plate III in the Appendix]. A work on the history of art by George Perrot and Charles Chipiez contains the following entry on the statue: "Le sculpteur a été moins heureux là où son attention n'était pas attirée soit par les masses musculaires très saillantes et nettement circonscrites. Les parties molles, telles que le ventre, l'ont fort embarrassé. C'est par un tracé tout arbitraire, par une ligne qui décrit une courbe en forme d'ancre renversée, qu'il a séparé de la cage thoracique la région abdominale." ("The sculptor was less happy when his attention wasn't attracted by the prominent muscle masses clearly circumscribed. The soft parts, such as the stomach, greatly embarrassed him. It is by a completely arbitrary route, by a line which describes a curve in the shape of an inverted anchor, which has separated the thoracic cage from the abdominal region.")

[*"Histoire de l'Art dans la Antiquité"*, George Perrot & Charles Chipiez, Volume VIII. La Grèce archaïque, p. 453,]

However, these peculiar kinks in the geometric grid simply follow the internal triangle of axes, intersected at the top by the upper horizontal diagonal of the "major". As this is an early example of the type of arrangement described above, this could be interpreted as mere uncertainty on the part of the sculptor to treat the torso in the context of the fundamental geometric form. Otherwise it is also possible that the statue simply remained unfinished.

The development of Greek sculpture in connection with the decagram can be traced from the stiff, archaic examples right through to the unfettered flux of the Hellenistic period. Much as was the case with the temples, there is a dual aspect to the compositions. Not only are the proportions of the human figure determined by the pentagram, but also the composition and flow of the entire artwork are based on this fundamental form. A remarkable example of this is the statue of Nemesis in the Vatican.

"Nemesis" in the Vatican. From Anton Springer, *Handbook of Art History*. The depiction is slightly askew because of the shifting of the axis.

Aphrodite of Melos [Venus of Milo] bends and flexes in the rhythm of the ascending diagonals [Plate XIII], while the "Bacchante" of Naples [Plate XIV] and the "Hellenistic Group" [Plate XV] are suffused with motion by the decagram. The best results are conveyed by those reliefs where an exact frontal aspect is not possible [Plates IV, VII, VIII].

Besides whole figures, the formation of the head is of particular importance. Plato speaks of a *"sphere-shaped body [...], which we now call the 'head', it being THE MOST DIVINE PART and reigning over all the parts within us."*

"τὰς μὲν δὴ θείας περιόδους δύο οὔσας, τὸ τοῦ παντὸς σχῆμα ἀπομιμησάμενοι περιφερὲς ὄν, εἰς σφαιροειδὲς σῶμα ἐνέδησαν, τοῦτο ὃ νῦν κεφαλὴν ἐπονομάζομεν, ὃ θειότατόν τέ ἐστιν

καὶ τῶν ἐν ἡμῖν πάντων δεσποτοῦν" *Timaeus* 44d

:

("The divine revolutions, which are two, they bound within a sphere-shaped body in imitation of the spherical form of the All, which body we now call the «head», it being the most divine part and reigning over all the parts within us." – R. G. Bury's Loeb translation.)

This statement makes clear that the face should not lie between the horizontal lines of the pentagons, but should be directly aligned with the reduced circle.

In the *Timaeus*, Plato goes on to say that "of the organs the gods first constructed light-bearing eyes", and this passage, too, is confirmed by magnificent examples. The head of the blonde *ephebe* [a youth aged between 18 and 20] in Athens is designed strictly according to the geometric net.

The large eyes are located at the focal points of the diagonals that emanate from the upper and lower axes, the heavy hair plunges onto the "Golden Section" of the radius, the curved lips follow two lines of the decagram.

The head of a caryatid on Sifnos. From *"Histoire de l'Art"* by Perrot & Chipiez.

The above example, a caryatid on Sifnos, is also very instructive, while the head of the Medusa Rondanini, a freestanding high relief, is of an exceptionally strict composition [Plate XII].

Compared to these works of art, which were created over the course of four or five centuries according to the principle of the decagram, their unbalanced predecessors and the decadent Roman artefacts can easily be identified [Plate XVI].

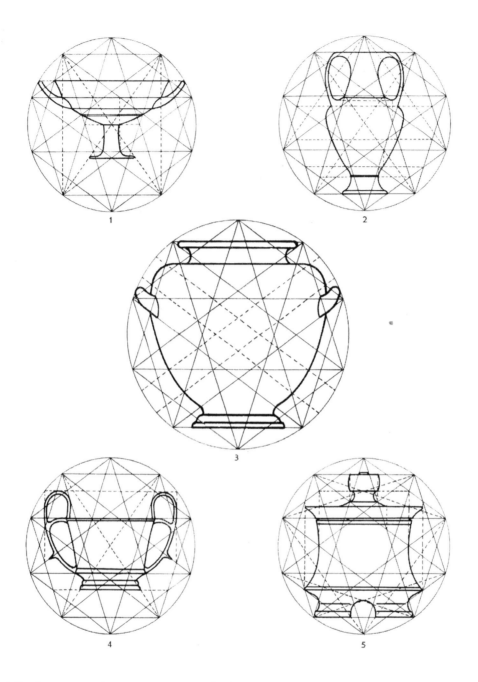

The images and measurements of the vases are based on *The Greek Vase* by Jay Hambidge.

Fig. 1 Kylix, Museum of Boston

Fig. 2 Amphora, Louvre, Paris

Fig. 3 Stamnos, Metropolitan Museum, New York

Fig. 4 Kantharos, Museum of Boston

Fig. 5 Pyxis, Museum of Boston

It remains to briefly survey the subject of Greek vases. They, too, owe the unsurpassed elegance of their forms to the laws of geometry. Some of their proportions are based on the ratios 1 : √2 or 1 : √3, but most are based, much like the temples, on the ratio 1 : √5, probably according to the proportions of their locations.

[Examples with the ratios 1 : √2 or 1 : √3 are excluded here.]

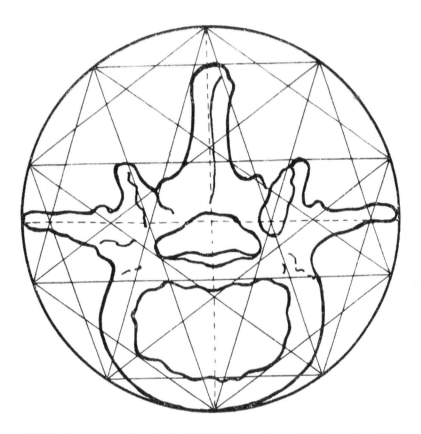

Third lumbar vertebra *(vertebra lumbalis III)*. From *"Anatomie"*, Werner Spalteholz.

III.

The consistency with which the Greeks applied their formal laws in architecture and sculpture over centuries suggests that they possessed a deep insight that served as an immutable foundation. In the *Phaedrus*, Plato puts the following words into Socrates's mouth: *"All great art demands discussion and high speculation about Nature."*

"πᾶσαι ὅσαι μεγάλαι τῶν τεχνῶν προσδέονται ἀδολεσχίας καὶ μετεωρολογίας φύσεως πέρι· τὸ γὰρ ὑψηλόνουν τοῦτο καὶ πάντῃ τελεσιουργὸν ἔοικεν ἐντεῦθέν ποθεν εἰσιέναι." Phaedrus, 270

("All the great arts require a subtle and speculative research into the law of nature …" –Henry Cary translation.)

("All the great arts require discussion and high speculation about the truths of nature …" - Benjamin Jowett translation.)

("All great arts demand discussion and high speculation about nature …" – Harold North Fowler translation.)

Both Luca de Pacioli (in the Renaissance period) and Adolf Zeising (around 1850) found evidence confirming the Golden Ratio in organic structures. Jay Hambidge, too, measured the human body and plants using his rectangles.

[*Editor's note:* The fundamental publication by Zeising, who was the man who invented the name "the Golden Section" (*Goldne Schnitt* in German), is Adolph (not Adolf) Zeising, *Der Goldne Schnitt*, Halle, 1884. Jay Hambidge (1867-1924) was author of many brilliant publications and was the editor of the periodical *The Diagonal* also. It was Hambidge who rediscovered the lost secrets of the Greek sculptors, which they learned in Egypt at the time of the 26th Dynasty, known as *dynamic symmetry*. He displayed these secrets geometrically by using rectangles, as has already been demonstrated above.]

Having obtained the above results in Greek art, our task now is to examine the natural proportions of the human body in relation to the double pentagram. Given Nature's so-called capriciousness, such an approach is likely to encounter considerable difficulties. Would small and large, strong and slim bodies in all their variations submit to a formal law? If this were indeed the case, there should be no exceptions even in the earliest stages of development or the subsequent major changes during maturation.

Here, such a new and wide field opens up before us that this investigation, with the help of but a few initial results, wants to point to one of the laws of Nature that Antiquity clearly recognised long before us, but kept secret.

In order to proceed with the greatest possible precision, the human skeleton was chosen as the subject of study. However, the mathematical image will remain relative. Irregularities caused by disturbances as well as external influences or resistances must always be taken into account when examining organic structures.

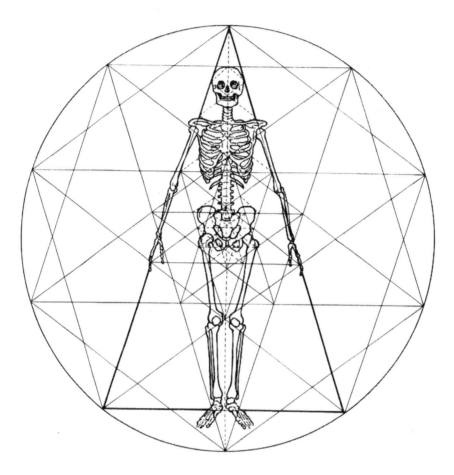

A standard skeleton.
From *"Anatomie"*, Werner Spalteholz.
(The position of the arms has been altered.)

The entire image of a human body aligns, much like Greek statues, within the double pentagon; the circle circumscribing the pentagon increases the total height by 1.236 ($\sqrt{5} - 1$). Head, axilla, sternum, pelvis and kneecaps align towards fundamental points. By slightly widening the arm position, it is much easier to integrate a human figure into the equilateral axis triangle (with its upper angle of 36°) than it is to integrate the figures of Leonardo da Vinci [Fig. 1] or Agrippa von Nettesheim [Fig. 2] with their outstretched limbs into a circle or a single pentagon.

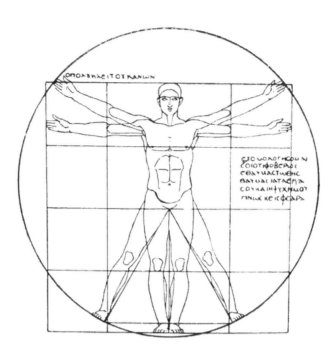

Fig. 1 Drawn by Leonardo da Vinci

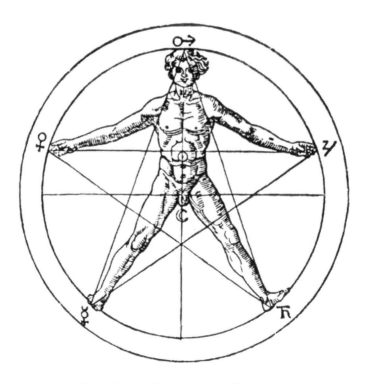

Fig. 2 Drawing by Agrippa von Nettesheim

The dimensions of individual bones can be verified on the basis of the geometric net of the entire figure. However, while the standard measurements of an entire skeleton are only reliable up to a point because the spinal column requires artificial layers [*discs*] between the vertebrae, the standard proportions of a skull can be determined with precision.

The study of the head – which, according to Plato, is most godlike – thus leads to further results. Skulls of a wide variety of shapes and races were examined. Compared to the study of the connections in architecture and sculpture, the task here is more complicated, even more so than with an entire skeleton. For not only the external image, the visible form, demands connection to the whole, but so does the internal bone structure. It seems that a two-dimensional fundamental figure is no longer sufficient and that further anthropological measurements are required.

The following illustrations should be understood as an extension of the application of the Greek formal principles to the human skull. They are only a preliminary stage to the final result and, in contrast to earlier considerations, do not refer to specific measurements and proportions, but to the perceptible conformity of the object with the angles and directions of axes and diagonals. The skeleton within the double pentagram [cf. the earlier illustration] already has revealed such relationships.

In the diverse illustrations below, the main focus are the protrusions at the zygomatic arch and the lower jaw towards the points of the pentagram. In the profile, where there are no symmetries that can offer any *a priori* points of reference for the geometric net, no clear overview – such as is the case in the *norma frontalis* – can be achieved. Here, the vertical axis falls between two points of the pentagram, as the orientation of the object at a right angle also requires the decagram to be rotated by 90°. The curvature of the skull aligns most closely with the fundamental form in the case of the European skull. The temporal bone reveals a certain relationship to the inner pentagon in all cases.

Norma frontalis

All examples of skulls are exact tracings of photographs printed in *Lehrbuch der Anthropologie (Textbook of Anthropology)* by Rudolph Martin

Fig. 1 Standard European
Fig. 2 Russian
Fig. 3 Swiss (from the Valley of Emmental)
Fig. 4 Chinese
Fig. 5 A child's skull

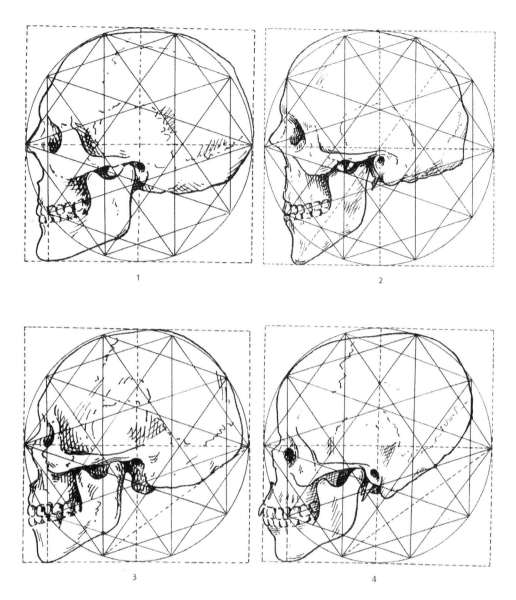

Norma lateralis

Fig. 1 European – profile of Figure 1, p. 49
Fig. 2 Swiss (Emmental) – profile of Figure 3, p. 49
Fig. 3 An inhabitant of Tierra del Fuego
Fig. 4 Negro

The most instructive view is from below and with the lower jaw removed (see following illustrations). The following is an attempt to apply the fundamental geometric net in different (but always proportional) sizes, in order to emphasize the regularity of the relationship between decagram and object.

The comparison between round and elongated skull shapes shows the almost unlimited possibilities of design.

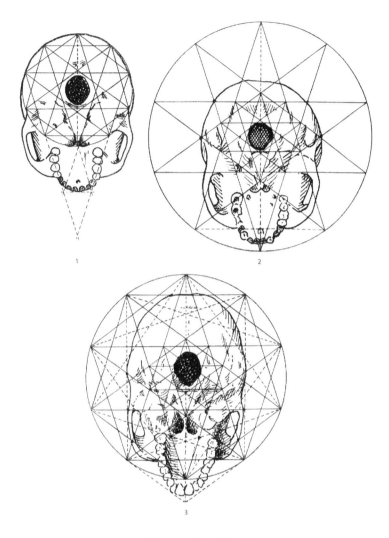

Norma basilaris

Fig. 1   Neolithic woman (Switzerland)
Fig. 2   Chinese
Fig. 3   European (this specimen is part of the author's collection)

Following the findings of the study of the bone structure, it would be useful to illustrate the formation of muscles and soft tissues. This is less distinct, but naturally conforms to the same principles.

At this point, anthropological measurements are needed to support the picture that emerged from the analyses of the illustrations. The standard measure of a skull is the distance from the *gnathion* (the lowest point of the midline of the lower jaw) to the *bregma* at the apex of the head. Further reference points are provided by the sutures and their multiple intersections, as well as maximal extensions, edges, ridges and apertures. After determining the overall height of the skull, it is instructive to take the principal relationships within the correspondingly large geometric net as a starting point, and to identify on the skull the radius, as well as the sides of the pentagon and decagon ("major" and "minor") and their various diagonals and intersections. Most often they can be located with surprising accuracy, although the exact positions of these segments is subject to some divergence depending on variations within the shape of the object.

For example, in a European oblong skull the following relationships apply within the geometric net:

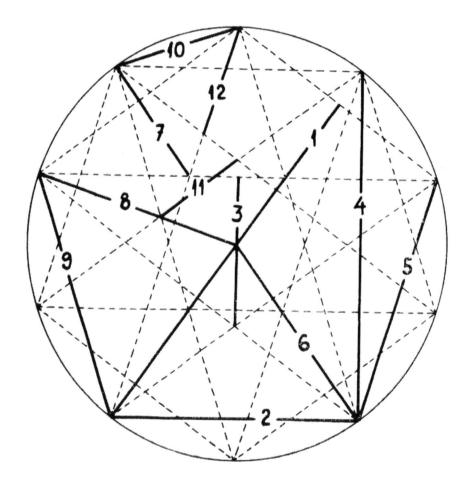

Height of skull (standard measure): 21cm

1. nasion – lambda
2. gnathion – nasion
3. prosthion – nasion
4. prosthion – bregma
5. zygion – zygion
6. mastoideale – mastoideale
7. inion – lambda
8. frontomalare temporale – frontomalare temporale
9. gnathion – lateral condylion
10. inion – asterion
11. mental protuberance – mental protuberance
12. infraorbital foramen – infraorbital foramen

Once again, Plato's words in the *Timaeus* are striking:

"He [God] fashioned a bony sphere round about the brain…"("καταχρώμενος δὴ τούτῳ περὶ μὲν τὸν ἐγκέφαλον αὐτοῦ σφαῖραν περιετόρνευσεν ὀστεῖνην, ταύτῃ δὲ στενὴν διέξοδον κατελείπετο") Timaeus, 73e and "designed our entire body around the life-giving marrow" ("…καὶ τὸν γόνιμον μεταξὺ λαβόντες μυελόν") Timaeus, 77d

Marrow and brain are described as the centre of the vital force, which radiates outwards in a state of equilibrium and creates the overall form. This proportion, which is perceptible as an, as it were, crystalline energy, is expressed especially clearly in the *norma basilaris* view of the skull, as well as the structure of the atlas and the dorsal vertebrae.

This analysis, carried out on the basis of the above measurements, can undoubtedly be considered successful. Within the fact that the human skull – in every aspect, in section and (from decile to decile) along the circumference – corresponds to a fundamental figure of appropriate size, the ultimate step back to the archetypal form is surely to be found.

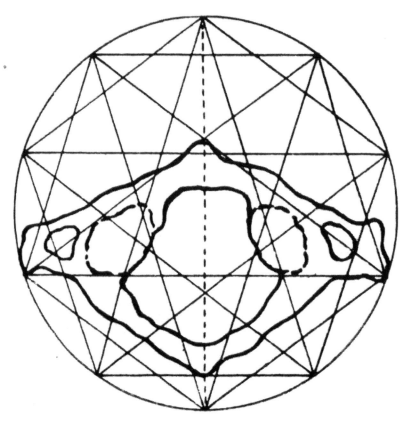

Atlas

When viewed from the back *(norma occipitalis)*, the sutures on the back of an Egyptian's skull (which, unusually, have not fused) reveal a REGULAR PENTAGON.

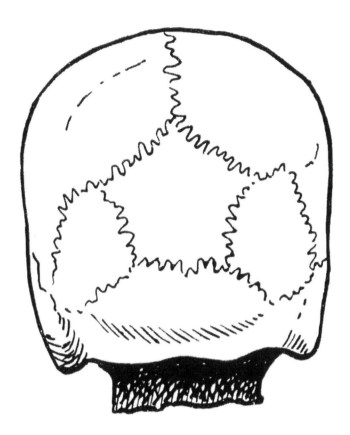

From *Anthropologie* by Rudolph Martin, No. 285, 1914
The regular pentagon in the centre is clearly seen.

Familiarity with this example makes it possible to recognise the pentagon even on skulls with a fused parietal bone. At adjacent surfaces – most obviously at the temporal bone – a DODECAHEDRON reveals itself as the theoretical structure for the shape of the skull. It is this solid figure that provides ultimate justification and explanation for the application of the pentagram in both nature and art. The pentagram, in its multiplicity and as the geometric principle of life and of organic development, corporealises the most perfect of the Platonic bodies. Most of the disparities that so far have marred the relationship between the norm and the natural object (or at least diminished the impression of congruence) can now be explained by the difference between the archetype and its two-dimensional plan that is applied in art. The displacements and inclination angles that emerge through the superimposition of pentagrams in the archetype will be the subject of future investigations.

Antiquity, personified in the figure of Pythagoras, derived its ideas of form and harmony with clear consciousness from the rich source that is knowledge of the pictorially lawful structures of Nature.

Archetype and image:
A small stellated dodecahedron and its silhouette

IV.

The dodecahedron, which Plato barely dares to mention, was the Pythagoreans' deepest secret. Within it, they recognized and venerated the revelation of the order of the world, both on a large and small scale. If we take this as a given, some of Plato's allusions that have so far remained puzzling can now be interpreted. The question whether Plato's *anima mundi*, the world-soul, is represented by a number or a geometric quantity was, according to Otto Apelt [the translator of Plato into German; 1845-1932], already discussed in Late Antiquity. In order to understand the attributes and supreme importance of the dodecahedron, let us evoke and examine its image once again.

Great icosahedron

The "intertwined" sides of twelve pentagrams extend outwards along the edges of an icosahedron. The twenty columns of the dodecahedron are each formed by three points of these pentagrams (alternatively, these can also be understood to constitute the great icosahedron), above which the dodecahedron's twelve regular pentagons are arranged. These, too, expand along their edges and form vertices, thus creating the supreme geometric figure, the small stellated dodecahedron. Plato's words, "concerning their generation into one another", thus find perfect expression.

"…περὶ τῆς εἰς ἄλληλα γενέσεως." *Timaeus*, 54d

('… all passed through one another into one another' in the Loeb translation by R. G. Bury, who edited the text to read differently in Greek; see not 54d but 54c in the Loeb edition.)

The icosahedron is the symbol of WATER (which, in a snowflake, crystallizes as a hexagon). From this fourth element of the inorganic domain, life metaphorically emanates through the dodecahedron and the twelvefold pentagram, flowing in and out perpetually, like the tides.

If motion in space is expressed through the circle that encompasses the regular geometric figure, then the pulsation of the solid figures can be equated with the flow of time. Ascent and decay, referred to in our modern era as the fourth dimension [*Time is widely considered as "the fourth dimension" in Einsteinian space-time physics.*], cause disruption in this constant flow. Contained within the dodecahedron are the cube, representing EARTH, and its constituent elements, the octahedron and tetrahedron [Plato constructs these using elemental triangles]. Together they achieve the harmonious synthesis of creation – "THE IMAGE OF THE ETERNAL GODS SUFFUSED WITH MOTION AND LIFE" (*Timaeus*, 37).

"For these reasons and out of these materials, such in kind and four in number, the body of the Cosmos was harmonized by proportion and brought into existence."

"καὶ διὰ ταῦτα ἔκ τε δὴ τούτων τοιούτων καὶ τὸν ἀριθμὸν τεττάρων τὸ τοῦ κόσμου σῶμα ἐγεννήθη δι' ἀναλογίας ὁμολογῆσαν" *Timaeus*, 32c

Pythagoras's teachings are based on measurements and numbers. Measurements and numbers are based on the small stellated dodecahedron.

And the artistic form of Antiquity is the – Platonic – shadow image thereof.

PART I: Archetype and Image

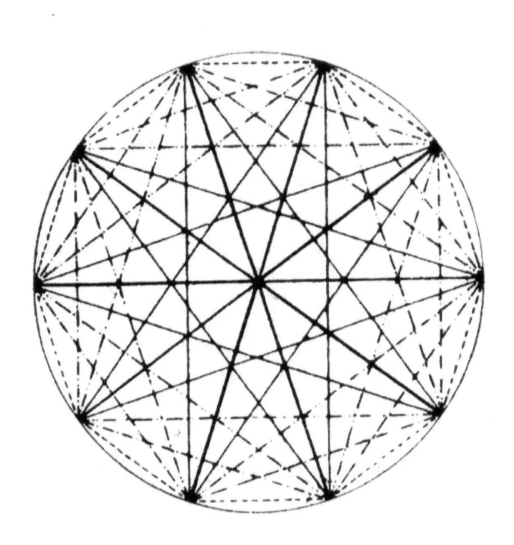

PLATES

A CARYATID ON THE ERECHTHEION
(London)

PLATE I

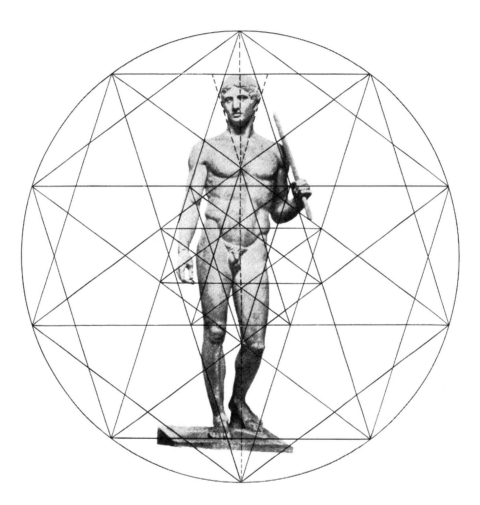

POLYKLEITOS: DORYPHOROS
(Naples)

PLATE II

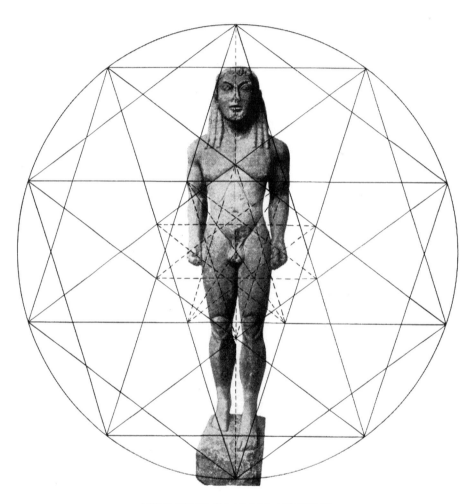

POLYMEDES OF ARGOS: ATHLETE
(Delphi)

PLATE III

DYING WARRIOR, SEPULCHRAL STELE
(Athens)

PLATE IV

PHIDIAS: ATHENA PARTHENOS (copy)
(Athens)

PLATE V

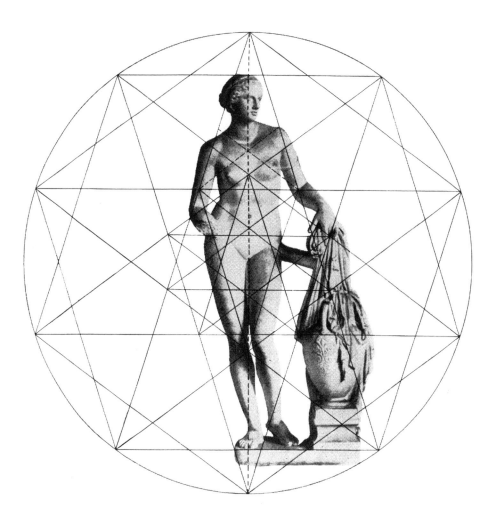

PRAXITELES: APHRODITE OF KNIDOS (copy)
(Vatican)

PLATE VI

A METOPE AT THE TEMPLE OF HERA IN SELINUS
(Palermo)

PLATE VII

THE SEPULCHRAL STELE OF HEGESO
(Athens)

PLATE VIII

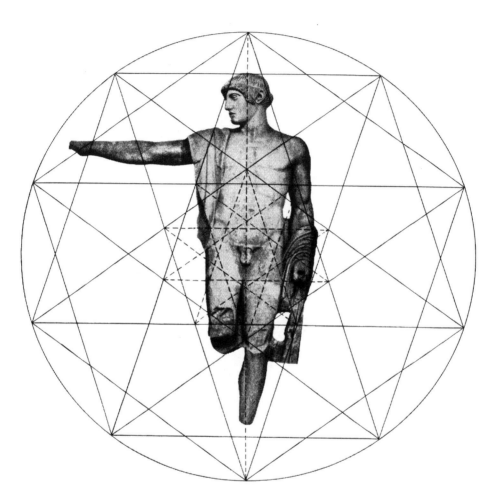

APOLLO ON THE PEDIMENT OF THE TEMPLE OF ZEUS
(Olympia)

PLATE IX

POLYHYMNIA
(Madrid)

LATERAL VIEW WITH OFFSET DECAGRAM

PLATE X

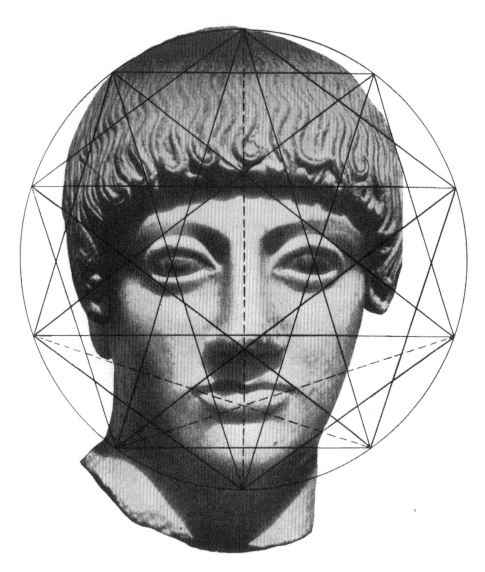

THE BLONDE EPHEBE'S HEAD
(Athens)

PLATE XI

MEDUSA RONDANINI
(Munich)

PLATE XII

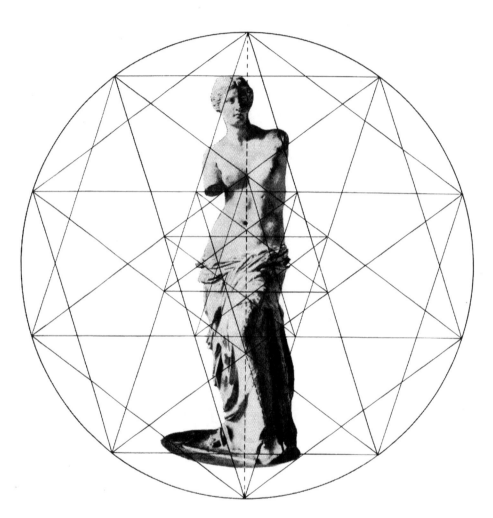

APHRODITE OF MELOS
[Venus de Milo]
(Paris)

PLATE XIII

BACCHANTE, section of a relief
(Naples)

PLATE XIV

HELLENISTIC GROUP
(Rome)

PLATE XV

1

2

3

4

COUNTEREXAMPLES
1 –3 Archaic, 4 Roman

PLATE XVI

BIBLIOGRAPHY

Bindel, Ernst   *Die Ägyptischen Pyramiden*, Verlag Waldorf Schule, Stuttgart, 1932

Boehn, Otto   *Von Geheimnisvollen Zahlen, Massen und Zeichen*, Verlag Sporn, Leipzig, 1928

Burckhardt, Jacob   *Griechische Kulturgeschichte*, Deutsche Verlagsanstalt Stuttgart, Berlin, Leipzig, 1930

Ghyka, Matila   *Esthétique des Proportions dans la Nature et dans les Arts*, Gallimard, Paris, 1927

Le Nombre d'Or, Gallimard, Paris, 1931

Hambidge, Jay   *The Diagonal* (1919-1920), *The Greek Vase* (1920), *The Parthenon* (1924), Yale University Press, USA

Martin, Rudolf   *Lehrbuch der Anthropologie*, Jena, 1914

Moessel, Ernst   *Die Proportion in Antike und Mittelalter*, Verlag C. H. Beck, Munich, 1926

Perrot , George & Chipiez, Charles

Histoire de l'Art dans l'Antiquité, Hachette & Cie., Paris (1883-1894)

Spalteholz, Werner   *Handatlas der Anatomie des Menschen*, Hirzel, Leipzig 1932

Springer, Anton   *Handbuch der Kunstgeschichte*, Seemann, Leipzig, 1901

Stenzel, Julius   *Zahl und Gestalt*, Teubner, Leipzig, Berlin, 1932

Theuer, Max   *Der Griechisch-Dorische Peripteraltempel*, Wasmuth, Berlin, 1916

Assistance with the illustrations: Ing. E. Leupp, Winterthur

Photographs: H. Wullschleger, Winterthur

PART I: Archetype and Image

PART II

THE GEOMETRIC PRINCIPLE OF THE GRECO-DORIC TEMPLES

PREFACE

The following study is an amateur's work, written after a journey through Sicily. I am deeply indebted to Professor Dr. E. Fiechter of the ETH Zurich (Swiss Federal Institute of Technology) for his scientific advice and assistance, and to Ing. E. Leupp for his help with the realisation of the architectural plans.

Winterthur, January 1939   L.W.-S.

PART II

CONTENTS

THE PLANS

Concordia Temple at Agrigento (Akragas)

I.

The fundamental idea of Greco-Doric temple building, and thus its symbolic function as a formative principle, had already withered and partly disappeared during the Hellenistic period.

Proof is provided by the Roman architecture that followed. While it is not entirely devoid of a geometric basis, it does fall far short of the equilibrium and poise of the Greek examples. Vitruvius's elusive lore has made it more difficult for posterity to solve the mystery.

There have been many attempts in recent time to rediscover these fundamental principles of Greek temple building.

This study came about as an attempt to provide as clear as possible a description – an, as it were, artisanal account – of the proportions of the Concordia Temple at Akragas. In the course of the work, guiding principles gradually emerged that relate to the entire topic.

Already at the end of the last century, Perrot and Chippiez's comprehensive work *Histoire de l'art dans l'antique* records various results in relation to the clarification of Greek proportions.[1]

The authors were probably the first to recognise certain metric proportions involving irrational numbers. Five examples of simple geometric figures lead to the determination of the perimeter of various temples. However, the following view is expressed about these important results: "*... que les operations de ce genre sont, en somme, plus spécieuses que probantes.*" ("*... operations of this kind are, on the whole, more specious than probative.*")

In another section of the work an attempt is made, using the so-called "*système modulaire*", to represent the main dimensions of the temple as multiples of a certain measure of length, the "*module*". This module corresponds to the dimension of any element of the building, such as the diameter of a column or the width of a triglyph. This type of approach is based on some allusions made by Vitruvius, which are not only rather superficial in and of themselves but are probably also interpreted incorrectly. The results are imprecise and unsatisfying.

---

1 Tome VII, Paris 1898, Chap. III, *"Le mode dorique"*.

II.

Two recent papers have provided valuable evidence for the geometric study of the Concordia Temple:

Die Proportion in Antike und Mittelalter (Proportion in Antiquity and the Middle Ages) by Ernst Moessel (Munich, 1926) and *The Parthenon and Other Greek Temples* by Jay Hambidge (Yale University Press, 1924).

While Moessel achieves mathematically precise results by inscribing the plans of particular edifices (both in horizontal and vertical projection) into a circle whose circumference is divided into ten parts, Hambidge uses his concept of "dynamic symmetry" to demonstrate how areas are clearly segmented according to the given measurements. The latter approach to design and analysis was a guiding principle at the outset of this study.

"Dynamic symmetry" determines the relationships between areas. These are the "root rectangles", the rectangle in the golden section (also known as "divine proportion"), and hybrid compound forms, using division and multiplication. For example, if a square with a side length of 1 and its diagonal $\sqrt{2}$ is used as a basis for a $\sqrt{2}$ rectangle, then the lengths of the edges of the rectangle will be $\sqrt{2}$ and 1 (1.4142 : 1).

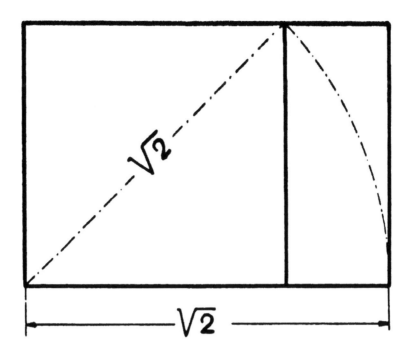

Figure 1: constructing a "root two rectangle" from a square

The base for the $\sqrt{5}$ rectangle is formed by two squares with the length 1 and a diagonal drawn through both rectangles with the length $\sqrt{5}$. [For readers who have never seen a swung diagonal of a square like this, it is important to mention that it is not an unusual phenomenon. Because the diagonal is known to have a length of $\sqrt{2}$, by swinging it down to the horizontal and treating the extra bit of extension created that way as turning the square into a rectangle, the long sides of the rectangle are certain of having a length of precisely $\sqrt{2}$. This eliminates the need to make calculations and measurements to achieve that length since the length becomes automatic. The resulting rectangle is then called "a root two rectangle", since the long sides have the lengths of the square root of 2. In one of the finest books ever written for non-mathematicians about geometry, Constance Reid in her book *A Long Way from Euclid* (Routledge, London, 1965) wrote: "Although no mechanical device can possibly mark off on a line the exact point which is the irrational distance from the beginning represented by $\sqrt{2}$, we can in theory mark off *the exact distance* by constructing a right triangle of unit size and swinging across the number line [the horizontal axis or baseline] an arc, the radius of which is the length of the hypotenuse, or $\sqrt{2}$. … The reader will note that for this "theoretical" construction of $\sqrt{2}$ we have used no measuring device like protractor or ruler.." (pp. 134–5) Reid chooses to swing the hypotenuse of a right triangle, which is the same as swinging the diagonal of a square divided into two equal triangles as seen in Figure 1. In both cases the length is $\sqrt{2}$.]

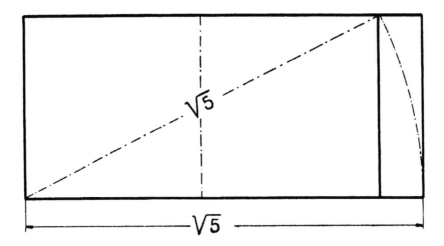

Figure 2: constructing a " root 5 rectangle" from a double square

The lengths of the resulting rectangle are √5 and 1 respectively (2.236 :1).

[This rectangle is called "a root five rectangle" since its long sides have the length of the square root of 5.]

A rectangle with the side lengths $\frac{\sqrt{5}+1}{2}$ and 1 is a rectangle in the "golden section" (1.618 : 1).

Hambidge's method for geometric analysis involves measuring architectural plans with these "dynamic" rectangles. For verification, the American method uses not explicit values, but rather their arithmetic calculation, the irrational numbers.

This paper will use the same approach.

The diagonals drawn in the various rectangles are of the utmost importance.

PART II

In his investigation of the Parthenon, Hambidge misses the fact that for the plans of temples his method of using rectangles in the golden section is unsuitable as it substantially complicates the metric arrangement. In addition, there are other reasons to reject his approach, as will be made clear in the course of this study.

III.

The Concordia Temple at Akragas has the following main dimensions:

Horizontal section, including euthynteria: length 42.23m, width 19.75m

Vertical section: height 14.30m, width 19.75m

The horizontal and vertical sections can be traced back to two basic geometric shapes, the square and the rectangle 1 : √5.

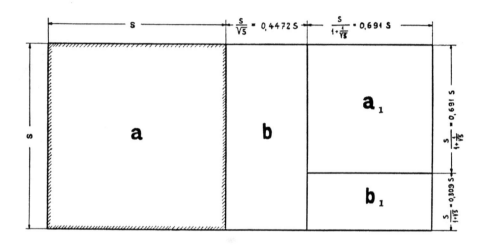

Figure 3.

Horizontal section:

$(a + b)$ and $(a_1 + b_1)$

are similar shapes.

Overall length:

$= 0.691s = 13.648m$

Verification of the dimensions:

$s = 19.75m$

$$\frac{s}{\sqrt{5}} = \frac{19.75}{2.236}$$

$= 8.832m$

$= 42.23m$

$$\frac{s}{1 + \sqrt{5}}$$

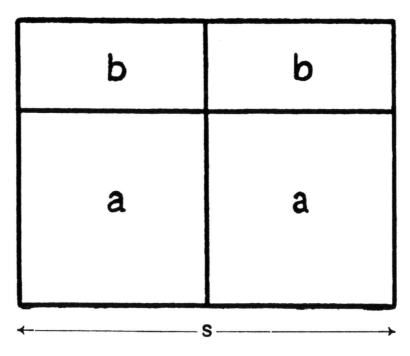

Figure 4

Vertical section: (a + b) stacked vertically form the two halves of the temple's width

Verification of the dimensions:

$$\text{Height} = \frac{s}{2} + \frac{s}{2\sqrt{5}} = \frac{19.75}{2} + \frac{19.75}{2 \times 2.236}$$

This unambiguous scheme of horizontal and vertical sections (Fig. 3 and Fig. 4) determines from the outset the continuation of the segmentation with only squares and rectangles $1 : \sqrt{5}$.

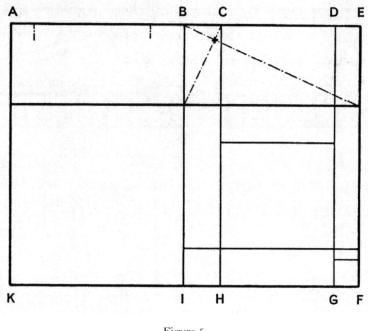

Figure 5

Figure 5 shows the development of the construction in the vertical section.

A diagonal through the rectangle 1 : √5 and a line drawn at a right angle determine result in a similar shape with the vertical line C H.

C D G H is a rectangle 1 : √5, D E F G consists of a square and four rectangles 1 : √5 stacked on top of each other. The square above I H gives the height of the stylobate, while the square below C D gives the height of the columns without the capital.

Figure 6 shows an elaborated scheme with the shape of the temple inscribed within. The dimensions of the individual architectural elements can be verified using the segmentation.

All of the temple's main dimensions are determined by the sides of squares and rectangles or by the intersections of their diagonals.

CONCORDIA TEMPLE AKRAGAS

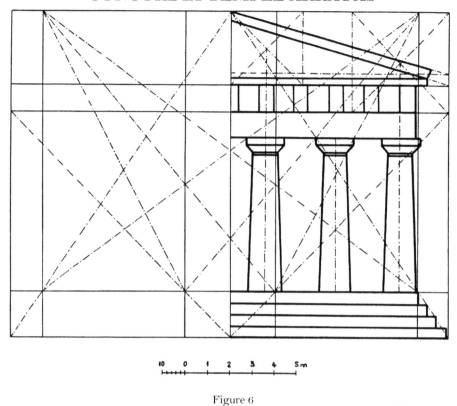

Figure 6

Example: the pediment including the frieze can be verified using the rectangle 1 : √5 (Fig. 4).

According to Serradifalco, the measurements are as follows:

Height of the pediment including geison and sima: = 3.21m

Height of the frieze: = 1.291m

 = 4.501m

Side of the rectangle 1 : √5 = 9.875m
Height of the rectangle 1 : √5 = 9.875m × 0.4472
(reciprocal value of 2.236) = 4.406m

Difference . . . . = 0.095m

Conversely, in the horizontal section, the schema of Fig. 3 allows only the determination of the width of the building's central section, using the small rectangle 1 : √5 (b₁). The segmentation does not yield results comparable to those of the vertical section.

An attempt to derive the proportions according to well-known theories relating to the stylobate, the bays separating the corner columns, the naos or the cella yielded several reference points, which were, however, more ambiguous than those obtained with the first method. It is especially striking how the powerful principal square in Fig. 3 loses its potency in this way.

The growing conviction that this element in particular must have an important function led to a new approach that had not previously appeared in the analysis, namely the combination of the schema, in its orthogonal segmentation according to Hambidge, with Moessel's circle and its section into ten parts. [Ernst Mössel, was author of *Die Proportion in Antike und Mittelalter* (*Proportion in Antiquity and the Middle Ages*), Munich, 1926, as mentioned at the beginning of this book.]

The result was the following:

<u>A circle, containing a double pentagram and inscribed into the principal square, reveals all the important dimensions of the temple, without their modification through division or multiplication.</u>

The key figure for all Doric temples of the canonical period has thus been determined. It governs the design in the horizontal section and at the same time incorporates the temple's vertical section.

Figure 7

The relationship between the double pentagram and the corresponding temple design is obvious.

CONCORDIA TEMPLE AKRAGAS

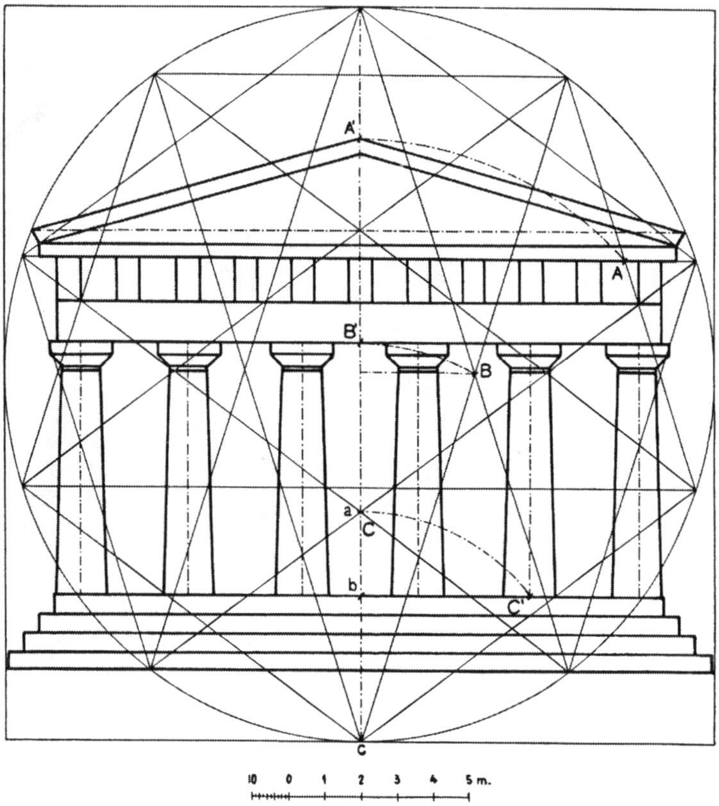

Figure 8. Vertical section in the circle with a double pentagram

However, it would be wrong to question the clarity of the system of rectangles as a consequence of this. Both geometric approaches lead to the same result.

The segmentation into rectangles has the advantage that it is possible to specify dimensions and proportions on the material and the edifice itself. Each separate geometric figure offers the opportunity for continual proportional segmentation in minute detail *in loco*.

CONCORDIA TEMPLE   AKRAGAS

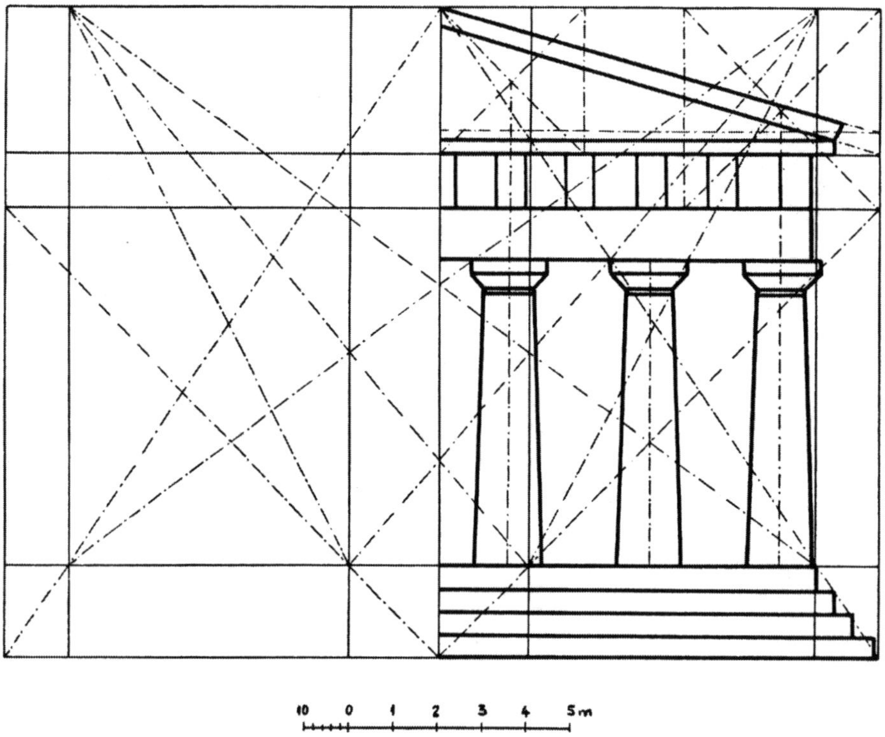

Figure 9. Vertical section in the rectangle.

It stands to reason that the key figure was perhaps a secret symbol that served as the fundamental principle for temple construction, while the rectangular arrangement was of public and practical use.

CONCORDIA TEMPLE   AKRAGAS

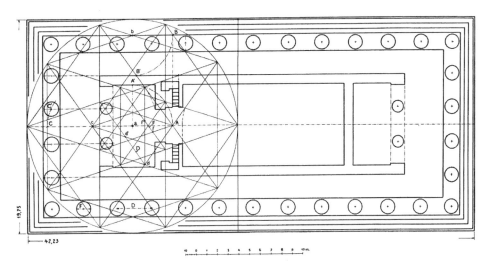

Figure 10. Horizontal section with key figure.

The distance between the arcs A A$_1$ and B B$_1$ provides the thickness of the walls in the central section; d d$_1$ = D provides the width of the intercolumnar bay along the longitudinal axis; and f f$_1$ = F provides the column diameter. The surface area of the cella is equal to 2 Q.

And yet there is an even deeper connection between the two conceptions. Their relationship to each other is based on the fundamental figure, the square with the double pentagram.

The properties of the double pentagram are well known. It owes its privileged position within the scheme to its sublime harmony pointing towards infinity.

It is a striking feature of the rectangular segmentation that the rectangle consisting of a combination of a square and a rectangle 1 : $\sqrt{5}$ (1 : 0.691) is prevalent everywhere and in every size. We have verified it in both the horizontal and vertical sections of the Concordia Temple; it can be identified in the horizontal section of the Parthenon; in the vertical section of the Temple of Aphaea on the island of Aigina, where it also appears four times in the central section including the step; it can be found five times in the cella of the Temple of Zeus at Olympia; and so forth.

In the square with inscribed circle and pentagon this form is determined by the intersection of two diagonals[2].

---

2 Hambidge: *The Greek Vase*, Chapter III, The Leaf.

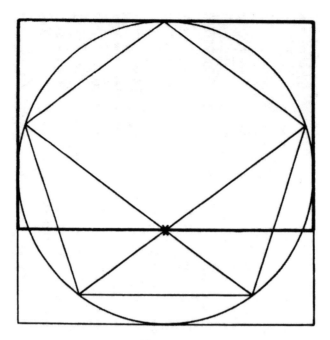

Figure 10.

In the following, it is this fundamental figure that will produce the other, most frequently used combinations of square and rectangle 1: $\sqrt{5}$. The Roman numerals accompanying the rectangles will reappear in the subsequent annotations of the temple plans, where the abbreviation Q will stand for "square" and R for "rectangle".

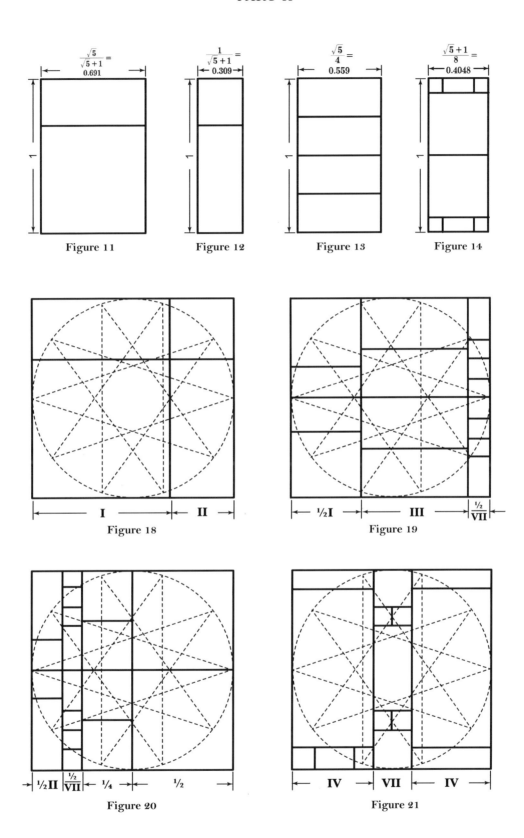

$$\frac{\sqrt{5}}{\sqrt{5}+1} = 0.691$$

Figure 11

$$\frac{1}{\sqrt{5}+1} = 0.309$$

Figure 12

$$\frac{\sqrt{5}}{4} = 0.559$$

Figure 13

$$\frac{\sqrt{5}+1}{8} = 0.4048$$

Figure 14

I   II

Figure 18

½I   III   ½ VII

Figure 19

½II ½VII ¼   ½

Figure 20

IV   VII   IV

Figure 21

Rectangle I is the continuously recurring form with symbolic meaning.

It is possible, by projecting the corners and intersection points of the double pentagram onto the sides of the square, to construct surfaces that consist exclusively of the two main figures of square and rectangle $1 : \sqrt{5}$. Therefore the shapes used in the orthogonal layout of the plan also can be found within the key figure; they are governed by it.

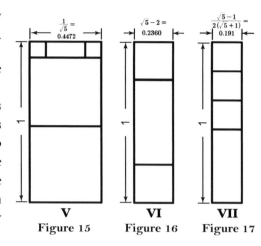

Figure 15   **Figure 16**   **Figure 17**

Consequently the results of the construction within the circle and within the rectangle are identical.

The four figures on the previous page (18-21) show the segmented surfaces at horizontal orientation of the points of the pentagram, while the diagrams on the right (22-23) show the same surfaces at vertical orientation of the points of the pentagram. The resulting rectangles are used in a variety of ways. The first type belongs primarily to the horizontal section, while the latter type appears in the vertical section.

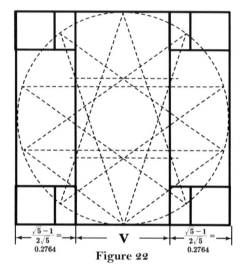

Figure 22

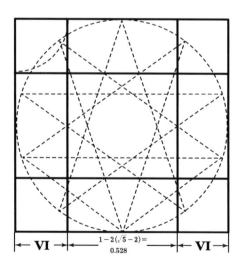

IV.

The following illustrations follow Max Theuer's plans (*"Der griechisch-dorische Peripteraltempel"*, *"The Greco-Doric Peripteral Temples"*, Berlin, 1918). The drawings were made exactly according to the data contained in that work, at a scale of 1:100. Small divergences in the mathematical verification are unavoidable for various reasons. For one, the poor structural state of some of the edifices makes precise measurements difficult. Secondly, until now it has only been possible to proceed in a reasonably systematic way by using Hambidge's method. The dimensions of the euthynteria (the uppermost course of the foundations), which are part of the overall plan, are usually omitted.

In the vertical section, the arrangement is derived from the key figure, with the lower horizontal side of the pentagon forming the basis. Where the pentagram's dimensions are not directly obtained, they are transferred onto the central vertical line or determined by that line. The points of the pentagram are always arranged vertically.

In the horizontal section, the points are arranged along the longitudinal axis of the temple. While the key figure cannot be established with certainty, it does govern the peripheral and central areas. The arrangement appears most clearly, as it does sin the vertical section, when seen from the side of the pentagon that is parallel to the front. The remaining areas adjacent to the square can be calculated through its division.

The dimensions of the intercolumnar bay and the column diameter can be found within the inner decagram.

TEMPLE C SELINUS

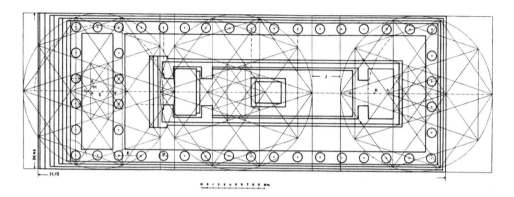

Figure 24

The horizontal section is composed of 2 Q + R I.
Length of the cella without rear step = 24.41m, width = 8.84m.
R I = (8.84 × 0.691 = 6.1084) is contained four times within the area of the cella.
Deviation of the mathematical verification: 0.023m.

Length E (e – e¹) = intercolumnar distance along the longitudinal axis
Length H (h – h¹) = column diameter

Length J = R VII,
Length K (k – k¹) = side length of the inner pentagon

TEMPLE OF APHAEA   AIGINA

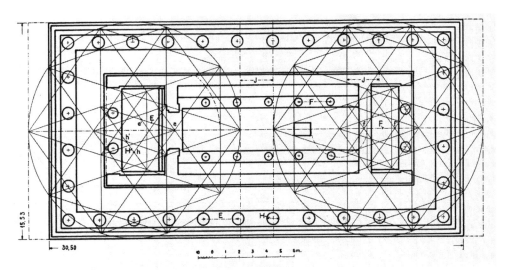

Figure 25. Segmentation according to the key figure

Calculation of the length using the key figure:

2×1 (side of the square) $- 0.191$ (short side of R VII) =   1.809
Length J = 0.309 (short side of R II) / 2   0.1545

          ———

          1.9635

width 15.53m $\times$ 1.9635 = length 30.493m

TEMPLE OF APHAEA   AIGINA

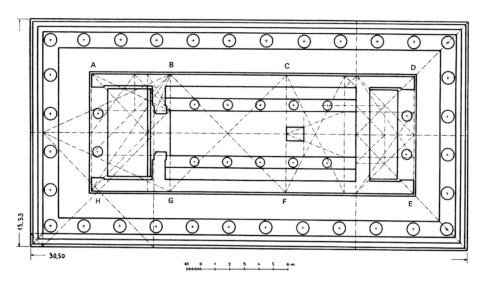

Figure 26 Orthogonal segmentation

The orthogonal segmentation in the horizontal section is anchored in the central structure. Dimensions of the walls excluding the step:

Length 22.60m, width 8.012m.

The composition of the shapes is clear:

A B G H = R I   A B = 5.536m
B C F G = Q   B C = 8.012m
C D E F = RV / 2   C D = 8.947m
                                               ⎯⎯⎯⎯⎯⎯⎯
(Difference 0.105m)   = 22.495m

The area of the cella (inside the pillars and entrance step) is 3.05m × 12.86m

It is composed of

2 Q, length of side   = 6.10m
R V, length of side 3.05 × 2.236   = 6.819m
                                               ⎯⎯⎯⎯⎯⎯⎯
(Difference 0.059)   = 12.919m

TEMPLE OF APHAEA   AIGINA

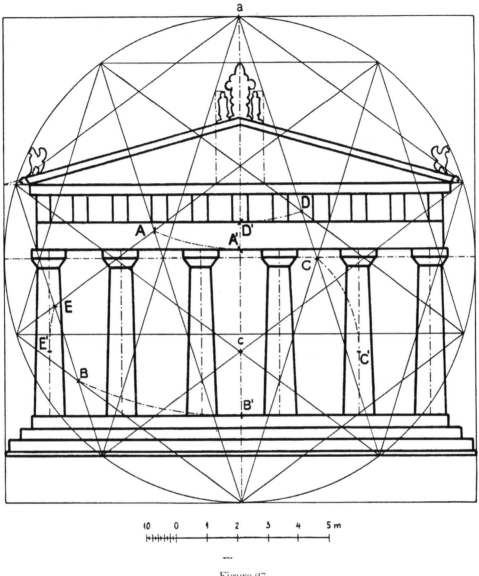

Figure 27

The reconstructed gable figures reach the upper side of the pentagon.

Vertical section Figure 29. The rectangle A D E H is composed of R V and Q = R I. Within the square, the vertical line C F determines the opposing R V. The further segmentation is derived from diagonals and the construction of squares in the lower parts of the rectangles.

The rectangle B C F G is composed of 2 R V and R I, and their shared side is the height of the supporting element.

TEMPLE OF APHAEA   AIGINA

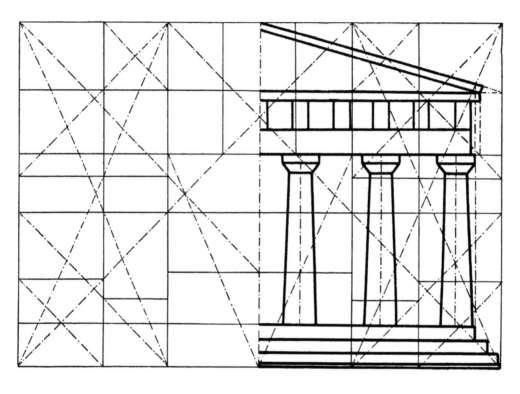

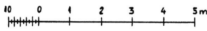

Figure 28

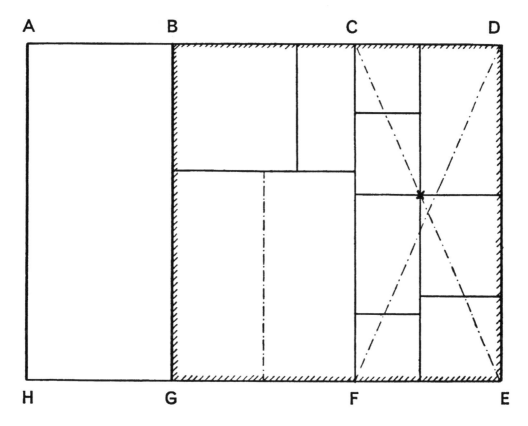

Figure 29

Verification of the measurements:

Total height 10.73m,
Width 15.53m

10.73m × 1.4472
(side lengths of Q and R V) = 15.528m

Difference = 0.012m

SO-CALLED BASILICA   PAESTUM

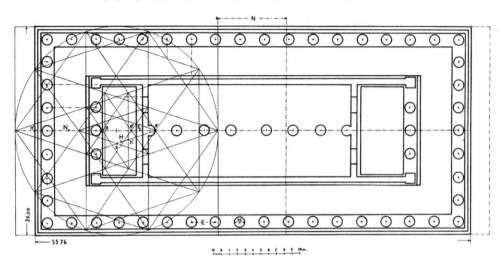

Figure 30

The horizontal section of the so-called Basilica at Paestum reveals a type of segmentation that deviates from the normal Doric type, which nevertheless is precisely governed by the decagram. The irregular arrangement of the central columns corresponds to Theuer's plan, which unfortunately does not contain the necessary measurements. The cella is a rectangle $1 : \sqrt{5}$. The width of the base of the columns in the stylobate is derived from a circle inscribed into the decagram.

TEMPLE OF ZEUS OLYMPIA

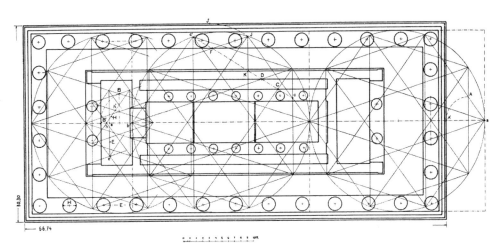

Figure 31

By way of exception, it is the *stylobate* that provides the basis of the key figure in this large-scale edifice. In order to show the cella's segmentation clearly, a decagram is superimposed over the centre of the plan. The line $c - c_1$ yields four important intersection points: C D K F.

The stylobate measures	24.68m × 64.12m	
It consists of $2 Q (27.68m \times 2.309)$ =	63.913m	
	(Difference = 0.207m)	

A B L M and D F G I = R V
B C K L and D C I K = R VI
E F G H $= 5 Q$

The two squares below E F govern the supporting element and the pediment. A square based on the stylobate and at the line D I yields the dimensions of the steps outside the rectangle A F G M.

TEMPLE OF ZEUS OLYMPIA

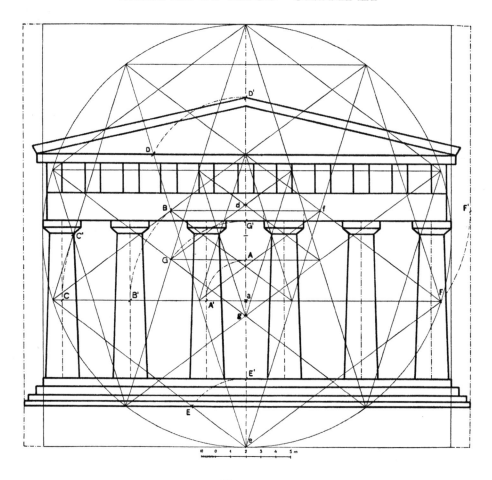

Figure 32

The height of the supporting element is equal to the side length of the inner pentagon projected from g onto the vertical line.

The building's height (20.27m) and width (27.68m) can be verified by way of calculation:

Length A M = 1

Length A B = 0.4472
Length B C = 0.236
Length C D = 0.236
Length C F = 0.4472

1.3664

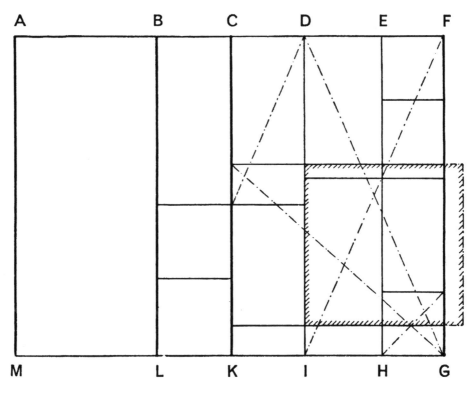

Figure 33

TEMPLE OF ZEUS   OLYMPIA

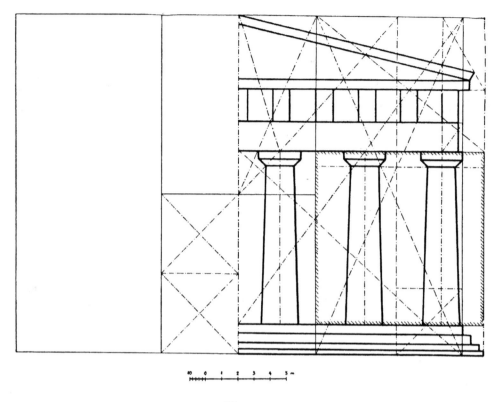

Figure 34

Height 20.27m × 1.3664 = width 27.696

Difference = 0.016m

The Temple of Zeus at Nemea is a building in the late Doric style. While an evolution of the forms in accordance with its period is in evidence – the columns have become higher and slimmer – the plan can still be derived with splendid simplicity and clarity from the key figure and rectangle.

TEMPLE OF ZEUS   NEMEA

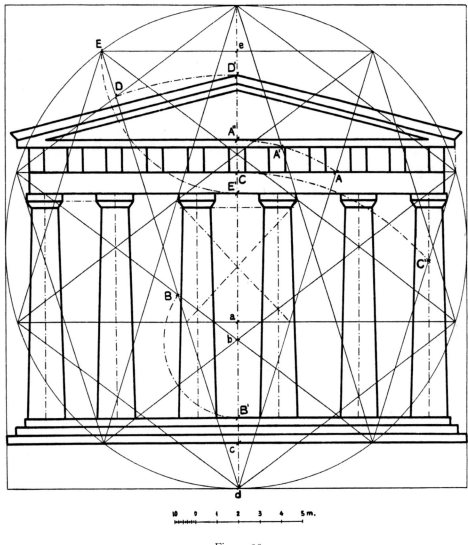

Figure 35

A B H I = R II
B D F H = Q

All proportions are determined by R I: A C M K, B D E L, B C G H = R I. The intersection point of the diagonals B E and C H yields the height of the supporting element.

Unfortunately, the absence of a full set of measurements makes any mathematical verification impossible. The height of the pediment can be determined without measurements by way of segmentation.

TEMPLE OF ZEUS   NEMEA

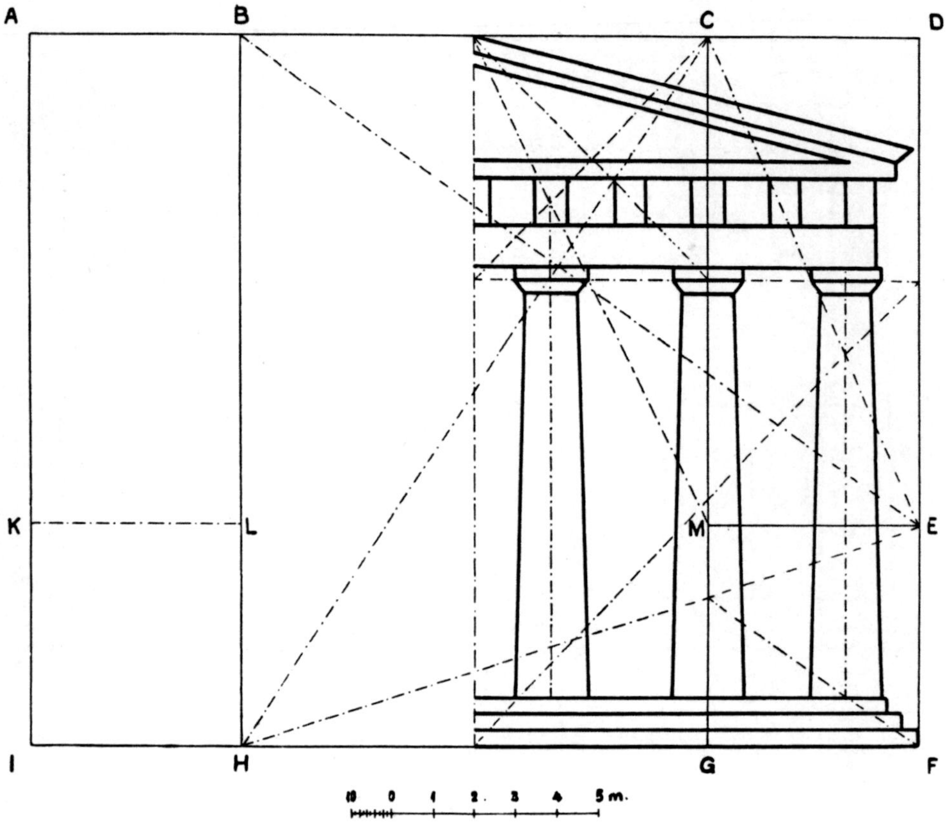

Figure 36

TEMPLE OF ZEUS   NEMEA

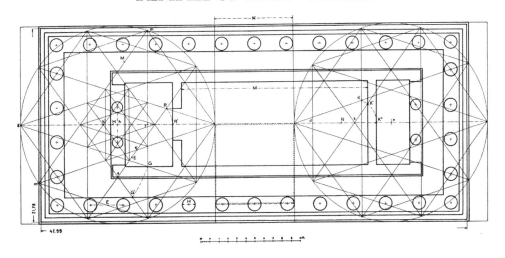

Figure 37

The horizontal section consists of $2 Q - R$ VII and R IV (centre). The central section of the building can be segmented into $1 Q$ and 3 R I. The cella is an R V.

V.

The plans of the six temples that have been examined in this paper, as well as those of the Parthenon and the Temple of Apollo at Bassae, reveal a consistent system of geometrical construction.

From this observation, the fundamental creative idea of temple building can now be inferred. The circle with the double pentagram represents the most sublime principle of the universe, all that exists endlessly, all that is unchanging, all that is indivisible[3]. It conveys its perfection to the square, the symbol of matter and becoming. Their union gives birth to "a third kind, which is always that of space"[4], namely the rectangle $1 : \sqrt{5}$.

One manifestation of this third fundamental figure is the oblong horizontal section of a temple, where the three elements of circle, square and rectangle are mutually interdependent.

As paragons of the divine, the Doric temples embody the trinity of antiquity.

This insight finally makes it possible to understand the widespread use of the rectangle I in the orthogonal organisation of the plans. Its shape encapsulates the image of divine trinity and thus it becomes the very symbol and determinative figure of Greek sacred architecture.

---

3 The principle of indivisibility corresponds to the principle of incommensurability in geometry.
4 Plato, *Timaeus*, 52b

VI.

Following the examination of temples of the canonical period, an investigation into the origin and decline of the proven geometric system would be of great interest. This topic can only be touched upon here, however.

After the archaic megara, whose dimensions were developed according to static requirements, it is likely that the first buildings were designed on the basis of the key figures of square, circle and octagon ($1 : \sqrt{2}$). An example of such proportions is provided by the horizontal section of the Megaron of Demeter at Gaggera / Selinus. Due to a lack of precise measurements it is not yet possible to provide an illustration.

The decline of Hellenic culture also brings with it a move away from the fixed Doric design principle. It remains uncertain whether during the course of the 4th century BC new, but less pure principles of architecture came to the fore. In any case, the practical system of rectangles outlived the circle with the decagram. Yet even its coherence seems to have become fragmented and lost. The lengths of the square, which had previously been determined by the double pentagram, could in the end only be understood theoretically, as units or *"moduli"*, and as such used to determine the dimension of building components.

The words of Vitruvius can be applied to this defunct remnant of Greek architecture: "The module is a gauge derived from the elements of the work itself."

What remains is the significant task of extending the results obtained here to the Ionic temples, to compare them on the broadest possible basis, and to interpret them comprehensively in architectural and cultural terms.

APPENDIX

Passage from Plato's *Timaeus*:

(Translation from Greek to German by W. F. Wagner, Breslau [Wrocław], 1841)[5]

"Yet it is not possible that two things alone should be conjoined without a third, for in the middle between the two there has to exist an intermediary bond. And the fairest of bonds is that which most perfectly unites into one both itself and the things that it binds together; and to bring this about in the fairest manner is the natural property of proportion. For whenever the middle term of any three numbers, measurements or forces is related to the last term in the same way that the first term is related to the last, and, conversely, as the last term is to the middle term and the middle term to the first, then the middle term becomes in turn the first and the last, while the first and last in turn become middle terms, and the necessary consequence will be that all the terms are interchangeable, and being interchangeable they all form a unity." (31c, 32a , 32b)

"Out of that which is indivisible and immutable and that which is divisible in bodies, God compounded a third kind of substance, midway in nature between the immutable and the Other, and thus placed it midway between that which is indivisible and that which is divisible in bodies. He took the three of them, and blended them all together into one form, by forcing the Other into union with the Same, although they are inherently difficult to mix. And after He had conjoined them with the substance and had made one out of three, he proceeded to separate the whole again into as many parts as were necessary, but in a manner that every part contained a combination of the immutable, the Other and the substance." (35b, 35c)

"However, He (God) gave sovereignty to the revolution of the immutable, to that which remains the same…" (36d)

"… the cycle of that which remains the same…" (37c)

"… these distances, intermediary- and connecting elements being not wholly dissoluble save by Him who had bound them together…" (43d)

"Let this, then, according to my verdict, be a reasoned account of the matter summarily stated – that being and place and becoming were existing, as three distinct things, even before the Heaven came into existence." (52d)

"… the becoming out of and into one another…" (54d)

"To Earth (matter) we assign the cubic form." (55e)

---

5 Of all the various translations of *Timaeus*, Wagner's is most useful for its references to geometric symbols.

PART III

PYTHAGOREAN TONES

AS RELATED TO GREEK SCULPTURE AND ARCHITECTURE

Printed as a first draft manuscript
for private and preliminary circulation to colleagues

PREFACE

This manuscript entitled "NOTE SYSTEMS AND PYTHAGORAS" for the moment is the last in a series of surveys on the foundations of Greek art.

It transfers the principles that can be derived from architecture and sculpture to the music of Antiquity. Before being finalised for general publication, it should be further clarified and extended by musicians. Its interim purpose is to record some results, to encourage further research into sound, and perhaps even to "set a new tone".

December 1942

Lucie Wolfer-Sulzer

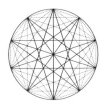

"NOTE SYSTEMS AND PYTHAGORAS"

This study is concerned with the intervals and combinations of notes that emerge from the Pythagorean proportions. It requires an alteration of our usual musical scales and a reinterpretation of existing knowledge about music in Antiquity. The following results were determined according to guiding principles that can be shown to form the basis of Greek art, especially architecture[1]. At the beginning of our endeavour it will be necessary to reject modern music theory, with the exception of a few absolutely fixed principles.

In Antiquity, music was placed above all other art forms, while architecture, surprisingly, was accorded much less importance. However, since every classical temple was based on "archetypal forms", such paradigms are bound also to manifest themselves in the music of Antiquity in some way.

In the following it is taken as a given that the Greeks did not only calculate their music, but also considered it within the geometric image, before bringing the two into harmony. Legend has it that Pythagoras found the most profound actualisation of his insights on the monochord, on a string. No reliable records of his original system have been directly passed down to us, and it is only with Plato that we can perceive comprehensive knowledge of the Pythagorean theories, which he must have acquired during his repeated visits to Sicily. Nevertheless, he appears to have taken a vow of secrecy[2], as his writings contain only allusions that are well-nigh impossible to decipher without a key.

In Plato's work, the harmony of the spheres includes that of the tones [notes]. Otto Apelt (a very fine translator) [*1845-1932, translator into German and commentator on Plato's works*] believes that Plato's description of the planetary system in the *Timaeus* is also relevant for the organisation of music in Antiquity. However, it is much more difficult to form a picture of the ancient organisation of sound on the basis of the brief mathematical terms contained

---

1 According to research by the author published in *Urbild und Abbild der Griechischen Form* (*Archetype and Image*), Fretz & Wasmuth, Zurich, 1941. The English translation of this work is Part One of this volume.

2 From *Le Nombre d'Or* (*The Golden Number*), by Matila C. Ghyka. Published by Gallimard, Paris, 1931.

therein than from another passage in Plato's writings, namely the dialogue between Socrates and Protarchos in his *Philebus* (25B – 26D)[3].

Let me begin by giving the relevant section of the dialogue:

Socrates: *"Were we not speaking just now of hotter and colder? Add to them drier, wetter, more, less, swifter, slower, greater, smaller, and all that in the preceding argument we placed under the unity of more and less."*

Protarchos: *"In the class of the* infinite, *you mean?"*

Socrates: *"Yes; and now mingle this with the other."*

Protarchos: *"What is the other?"*

Socrates: *"The class of the* finite *which we ought to have brought together as we did the infinite; but, perhaps, it will come to the same thing if we do so now; when the two are combined, a third will appear."*

Protarchos: *"What do you mean by the class of the finite?"*

Socrates: *"The class of the* equal *and the* double, *and any class which puts an end to difference and opposition, and, by introducing number, creates* symmetry *and* harmony *among the different elements."*

Protarchos: *"I understand; you seem to me to mean that the various opposites, when you mingle with them the class of the finite, take certain* forms.*"*

Socrates: *"Yes, that is my meaning."*

Protarchos: *"Proceed."*

Socrates: *"Does not the right combination of opposites give health in disease, for instance? And whereas the high and low, the swift and the slow are infinite or unlimited, does not the addition of the principles aforesaid introduce a limit, and perfect the whole frame of music?"*

This clear reference by Plato to the organisation of music deserves the utmost attention. He has Socrates bring into *harmony* and *symmetry* the two opposed classes of the *infinite* and the *finite*. The *infinite* is associated with the unlimited and the eternal, while the *finite* refers to matter and all that is transient. In the *Timaeus* (36C), Plato describes the same concepts as *"fraction"* (the result of irrational relationships, especially those in the pentagon) and *"number"* respectively. [*In other words, Plato classed irrational numbers as "infinite"*

---

3   Plato also talks of music in his *Laws* and *Republic.*

because they can never be fully computed; in modern times, they are expressed as endless decimals, but in ancient times they were expressed as an endless series of fractions. Numbers which are not irrational were considered by Plato to be "finite".]

Through the geometric image of the double pentagon with its diagonals, the words of Socrates become intelligible and infused with meaning. This figure is the mysterious basis of classical Greek art. It is the projection, the "image", of the fifth Platonic solid, the dodecahedron, when its sides are extended to form a stellated dodecahedron.

Figure 1
Stellated dodecahedron,
the "archetype"

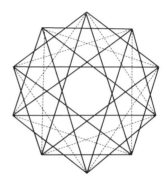

Figure 2
Double pentagon with diagonals,
the "image"

The simple pentagon inscribed in a circle, divided according to the Golden Section, constitutes the precursor of the composite figures (Figure 1 and Figure 2), that is to say, its *doubling* and *completion* in the decagon.

If, according to legend, the *monochord* of Pythagoras is connected to this symbol, the key to the universe of Antiquity, then an investigation of the notes and the relationships of their intervals will become possible most straightforwardly on a *string*.

A string is of a *given length*. Its whole length produces the fundamental note; shortening it will raise the resultant note. The frequencies of the various notes are in exact inverse proportion to the length of the string. From a physical perspective, we will have to accept the following natural facts:

1. The ratio of the length of the string from octave to octave = 2 : 1

2. The relationship of the fundamental note to its (resonating) overtones, that is to say, to the first octave, the second octave plus a fifth, and the third octave plus a fifth and a third.

Length of the string
Figure 3

The first octave concludes exactly at the halfway point, the second octave at the following quarter, the third octave at the following eighth, etc.

If a string of given length is placed around a circle with an inscribed pentagram and the resultant segmentation (as shown in the following diagram, Figure 4), then the above octave relationships will emerge automatically. In the triangle *OTU*, each (successively smaller) pentagon that adjoins the central pentagon is exactly half the height of the previous one. It must come as a surprise how easily the concepts of *doubling* or *limiting* integrate with the geometric shape of the pentagon, the epitome of the irrational relationships in the constant proportion (or Golden Section).

[*This is the key to the author's astounding insight into how the musical tones could be defined by a geometrical diagram based on pentagons, thus demonstrating that the tones follow the same geometrical schema as Greek architecture and Greek art, and also incorporate the Golden Section at the very heart of harmonics.*]

AO is the length of the string, which corresponds to the fundamental note; *B* is the centre point of the string; *AB* is the first octave; the length *BC*, as the second octave, is half of *AB*; *CD* is half of *BC*; much as *DE* is half of *CD* etc.

The completion of every octave pentagon occurs through the division into ten parts of the surrounding circle.

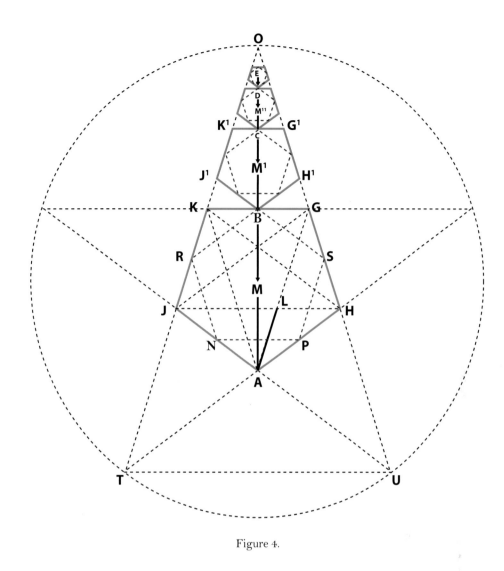

Figure 4.

Two pentagons taken together – with the smaller shape (*NPSBR*) [*shown in red*] resulting from the halving of the sides of the larger shape (*AHGKJ*) [*shown in blue*] – by turns bring into being the sequence; that is to say, the extension of the side of the inscribed pentagon (*RB* and *SB*) generates the side of the above, circumscribed pentagon[4].

However, the geometric schema reveals much more. The relationship between the fundamental note and the overtones is connected directly to the octaves, through the side of the pentagon *AJ* and a segment of the diagonal,

---

4 The Egyptian letter "mas" provides an insight into the symbolism of Antiquity. "Mas" means "birth" (cf. the journal *Vie, Art, Cité*, Vol. 2, Lausanne, 1942). The symbol can be understood as a fragment of the image of the tapering of the pentagons (Figure 4). Pythagoras visited Egypt [as did Plato at a later date].

AL[5]. On the string *AO* these lengths appear in reverse, that is to say, the whole string, reduced by the length *AJ*, sounds as a fifth to the fundamental note, and, reduced by the length *AL*, sounds as a third to the fundamental note. While the relationship of the third is, as it were, exact by our modern standard (being located somewhere between the "original" third and the later, tempered third), the fifth sounds lower than it does in the modern tuning system (string ratio 0.6708 instead of 0.6666). Another natural overtone (of the note c) is b flat" in the fourth octave. The string length that corresponds to this note (unlike that of fifth and third) cannot be derived from the pentagon. However, if the pentagon is expanded into a *double pentagon*, then the tenth note of the overtone series does lie on the decagon[6].

The various string lengths can easily be derived from this figure:

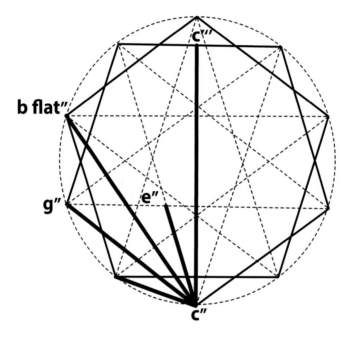

Figure 5.

The harmonious arrangement of the overtone lines is remarkable. They radiate in angles of 18° from the vertical axis and are complemented by the *side of the decagon*, which corresponds to a note whose pitch is between e flat and e. The symmetrical image of the relationships between the overtones thus appears as a schematic plant:

---

5 The diagonal of the pentagon is divided according to the Golden Section. Taken together, the first two segments are equal to the side of the pentagon.

6 The 10 notes of the overtone series:

1	2	3	4	5	6	7	8	9	10
C	c	G	c'	e'	g'	b flat"	c"	d"	e"

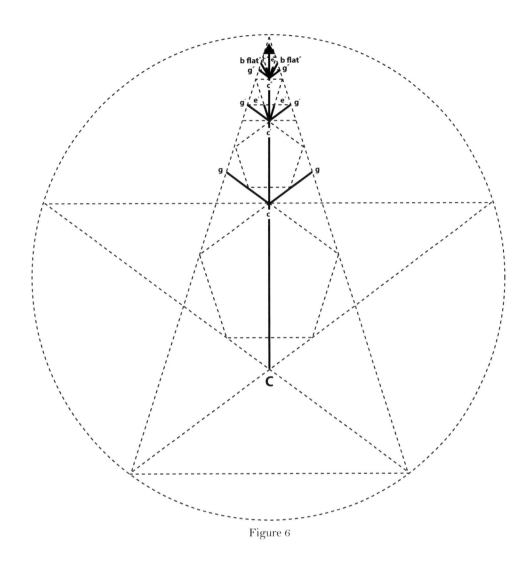

Figure 6

Note: Similar proportions can be found in the structure of the human body, as in patterns created by organic development in general. In basing the structure of the ear and the structure of music on the same geometric concepts, Pythagorean doctrine allows for the emergence of full CONGRUENCE.

However, this clear harmonic arrangement brings us into conflict with an important musical principle. It is nowadays considered established that only *simple relationships between notes lead to the best consonances.* The interval of the fifth is based on an exact trisection of the string, the interval of the third on a ratio of 1 : 5; and all the other intervals of our scales are derived and calculated from *regular* string segments. However, if the intervals between the notes are derived from the pentagon, then the basis for all relationships and divisions is fundamentally altered; they take on the characteristics of the irrational or *unlimited* in Platos's dialogue.

It is therefore not appropriate to define or regard, from our modern point of view, the Greek scales as *correct*. Moreover, such interpretation has never yielded any satisfactory musical results.

The question of the interval of the fifth is especially important. In his book *Science and Music*, Sir James Jeans, a professor at the University of Cambridge, discusses it and the discrepancies connected to it in great detail. Starting from the note c, he arranges twelve consecutive fifths around a circle, much like a face of a clock, and demonstrates that twelve fifths do not add up to seven octaves. Arranged in ascending order, our modern interval of a fifth ($^2/_3$ of the string length) is equivalent to a successive multiplication by 1.5.

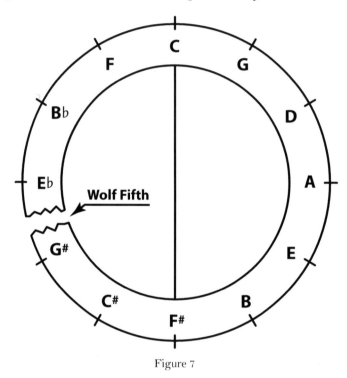

Figure 7

From Sir James Jeans, *Science and Music*, Cambridge University Press, 1937. [*Editor's note*: Wolfer-Sulzer calls the Wolf Interval by the name of *Wolfsquinte*, "Wolf Fifth". By doing so, she was thinking of how a fifth can become horribly out of tune if it is played with the addition of a tiny frequency interval, so that instead of a perfect frequency ratio of 2 to 3, it is slightly more. This can be avoided by special tuning of the strings. When such a discord happens, musicians called it "howling" like a wolf. Hence the name "wolf". But Jeans himself in his diagram did not say "Wolf Fifth", he says "Wolf Interval". So it is necessary to make this terminological correction here. The Jeans diagram does not depict a single fifth, it depicts a spiral (shown flat on the page) of twelve ascending fifths. The arithmetical frequency discrepancy found when adding up the frequencies of twelve fifths and comparing it with what you get by adding up the frequencies of seven octaves (climbing up two separate sets of stairs side by side, you might say) is a tiny fraction which we can show as a ratio in decimal form as 1.0136. So one is larger than the other by 0.0136. (The ancients never wrote it as a decimal, they always wrote it as a fraction. Indeed, the most precise value given in an ancient Pythagorean text known as the *Katatomē Kanonos* is the ratio of "531,441 to twice the value of 262,144", which computes out as a decimal of nine decimal points with perfect accuracy to the accepted modern value of 1.013643265. This is one of the most astonishing mathematical feats of antiquity, but is essentially unknown by historians of science and mathematics.) That tiny interval is very famous and is known as the Comma of Pythagoras, named after the ancient Greek philosopher who was the first known Greek to incorporate it into his theory of music and harmony. So the diagram of Jeans is really a way of showing the Comma of Pythagoras as a (flattened) spiral which falls short by a tiny gap. The numbers defining the Comma of Pythagoras are as follows: seven octaves arrive at a frequency 128 times higher than the first note, but twelve fifths arrive at a frequency 129.75 times higher than the first note. 129.75 divided by 128 gives the number 1.0136. This tiny discrepancy may seem insignificant, but it is not. The Jeans diagram is somewhat misleading, as it is the ascending spiral of octaves (which is not shown by him) which falls short, not that of the fifths (which he shows). And Wolfer-Sulzer was led astray by Jeans's own confusion. As her life was cut short, and the booklets which constitute Part III and Part IV of this book were both circulated privately as draft texts, which she admitted were incomplete, she did not live long enough to achieve full clarity regarding the Comma of Pythagoras, which in fact she does not mention.]

Figure 8. Sir James Jeans (1877–1946), author of *Science and Music*.
[Jeans was a famous astrophysicist, mathematician, astronomer, physicist, and musician, "one who could truly be described as a Renaissance man", according to one biographer.]

However, in practice, a successive arrangement of fifths (as if around a clock face) leads to a "wolf fifth", as it is called, which is a very dissonant interval.

The necessary consequence is a slight trimming back of the relationship of the fifth, in the same way that on string instruments, as is well known, the fifth has to be tuned "slightly lower". Jeans (p. 188) even provides the correct multiplier for the series: namely 1.49535 or $\sqrt[4]{5}$, instead of 1.5. In turn, the root of 5 and its properties are intimately connected to the pentagram and the dodecahedron.

On the circle of fifths, the notes c and f# are located on the vertical axis. Nevertheless, every interval is marginally too wide. F#, when assigned a lower pitch, in the geometric schema of a single octave is located in the centre of the pentagon (letter *M*, Figure 4). It should be noted that this centre point does not bisect the length of the octave into exact halves, but rather divides it in the ratio of 0.5528

$$\left(1\frac{-\sqrt{5}}{5}\right):0.4472\left(\frac{\sqrt{5}}{5}\right)$$

that is to say, almost as 5 : 4. The shorter, upper part *MB* is related to the longer, lower part *BM'* as 1 : 0.618 – in other words, the well known ratio of the Golden Section.

The *unlimited* (the irrational) and the *limited* (relationships of the octaves) continually depend on each other through these relationships and, according to Plato, *perfect the whole frame of music.*

The theory of the "simple relationships of the consonances" is now beginning to unravel before our eyes. However, we will not have to overthrow the entire system of our arrangement of notes; it is rather a re-evaluation of the divisions, which brings with it new and rich opportunities and will hopefully not be rejected out of hand by all sides. Our modern music will remain unaffected, and it should be mentioned at this point that the intervals of our modern scales can, surprisingly, also be found in the pentagon.

It is very tempting to carry out musical experiments on a primitive instrument (such as a wooden board with a large number of strings of equal length attached to it), by measuring certain intervals and trying to identify such chords that correspond to the relationships in the decagon. When struck, our familiar intervals arrange themselves around the *fundamental note* and its octaves. The result is a collection of notes that has been balanced long since and is linked to the half length of the string, as well as the concepts of the *number* and the *limited.* If this approach is contrasted with the Greek concepts of the *unlimited* or the *fraction*, one can conclude that the series of notes that has been lost must have emerged from the centre of the pentagon, which is simultaneously the centre of the circumscribing circle. The *centre* is always emphasised in Plato's philosophy. Ascending or descending from that point on the string (letter *M*, Figure 4), which, as mentioned earlier, in the C major scale corresponds to a note with a pitch somewhere between f and f#,[7] it is possible to find strange and beautiful sounds. They can be arranged in thirds in such a way that, within an octave, four lighter and four darker notes combine to form major - or minor - like harmonies. While the lighter notes emerge from the centre of the pentagon,[8] the darker notes, which lie in between, are none other than those derived from the harmonic series of the decagon (the sides of the decagon and pentagon and the diagonal of the decagon, Figure 5). The two series of notes (three and a half thirds within the octave) swap their functions as they continue; that is to say, until they repeat, they span two octaves, the ancient *diapason*. This results in a scale of seven notes, which corresponds to Greek musical tradition.

With very few exceptions, the kithara and the lyre are depicted ⌈in ancient art⌉ as having seven strings. The term *tetrachord*, which nowadays denotes a diatonic series of four notes within an octave, regains its original meaning of a *chord* struck on four strings. Furthermore, the terms *diapente* and *diatessaron* become intelligible when they are related to the length of an octave bisected by the centre of the pentagon (Figure 4: *AM* = *diapente* = "through the five"; *MB* = *diatessaron* = "through the four" of the nine parts). The much-mentioned *limma*, a remnant or vestige, must not be understood as the smallest of the intervals. Rather, it is an interstice with a certain shimmering quality that arises in the development of the intervals. In the *Timaeus*, Plato equates the *limma* with the numerical ratio 243 : 256, while in the geometric schema this proportion is expressed as the ratio between the height of the pentagon and its diagonal[9].

---

7   As the eleventh overtone also lies somewhere between f and f#, it probably also coincides with the centre of the pentagon and thereby represents a point of culmination within the Greek tonal space, which spans four and a half octaves.

8   The ratio of the string lengths:

$$I : 0.8944 \frac{2\sqrt{5}}{5}, \quad I : 0.7236 \frac{5+\sqrt{5}}{10}, \quad I : 0.6056 \frac{5+2(5-2\sqrt{5})}{10}$$

9   According to certain expositions in the fifth book of Euclid, the major relationships within the double pentagon can be coherently represented as a geometric series. [Editor's note: *Book Five of Euclid does not specifically mention pentagons, but the author may have worked out a way of using her 'double pentagons' described earlier in this work (i.e., one inscribed inside the other, one pointing up and one pointing down, shown in red and blue in Figure 4) to demonstrate for herself some of the comparisons of magnitudes to which Book Five is dedicated, with the various series of ratios and proportions, of which that Book contains a bewildering amount. Otherwise, she may have taken her notion from the discussions of pentagons in the latter part of Book Four, Propositions 11 to 14. So the representation by a geometric series must be the author's own creation, of which alas she has left no description, this booklet having been printed in an abbreviated form for private circulation to her colleagues. The most likely possibility seems, however, to be that the author derived her ideas from the Definitions of Book Five, especially Definition 11.*⌉ According to this view of the matter, the limma appears as the smallest progression.

[*Editor's note:* The author has come up with an important observation here, apparently not previously known. Few people understood clearly what the *limma*, written in Greek as *leimma*, is. It is a small interval but not the same as the Comma of Pythagoras mentioned earlier. One way to explain it is to suggest that it is a diatonic semitone. A good short description which I have encountered is that given by the musicologist Alain Daniélou in his book *Introduction to the Study of Musical Scales*, The India Society, London, 1943, p. 40: "… the limma is situated between the minor and major half tones …", p. 230: "There are, however, some intervals whose difference is so slight that the ear cannot easily detect it but whose functional difference is important: such is the case for the limma 256/243 …" "It is, therefore, a complementary interval in the ascending scale, but a direct interval in the descending one, as the ancient scale was.", p. 244. It is thus an exceedingly tiny musical interval, but its presence or absence makes a huge difference; after all, bees are small, but without them there is no honey.]

This first broad outline of the Pythagorean note system thus reconstructed does not yet allow us to embark on a reinterpretation of the musical scales whose names – Dorian, Lydian, Phrygian – have been preserved to this day. However, more clues can be found in the writings of the Roman writer Vitruvius, notwithstanding the fact that his reports do not go beyond superficial enumerations. It is significant that the fourth chapter of Book Five of his treatise *On Architecture* deals with "harmony". Yet, any profound insight into the ancient musical system had already long been obscured by the time of the Emperor Augustus, when Vitruvius was writing. It seems plausible that the Greek notes were organised into three "genera", namely *enharmonic, diatonic* and *chromatic*; that the octave was divided into 18 notes (as is evident through the surviving note series) and that steps could either be whole tones, half tones or quarter tones (*diesis*).

[*See the Editor's Note at the end of Part III, below, for more information about the Vitruvius passage.*]

As the "harmonic" tonal relationships can be derived from the larger pentagon of octaves, it seems possible that the inscribed pentagon (*NPSBR*, seen *in red* in Figure 4), which generates the shorter series of the form, also has a certain significance. If it is completed *in itself* into a decagon or double pentagon, a wholly unexpected arrangement emerges: the horizontal diagonals of the figure divide the axis – into the intervals of *our conventional tonal system!* Although the intervals in the two approaches are derived using completely different mathematical methods, they are ultimately so similar that the resulting sonic effect is the same. Through the tempering of our modern scales a slight displacement towards the irrational relationships has already occurred, as can readily be seen in the following table:

STRING LENGTHS

			tempered	relationship in the pentagon
c − d	0.8888	8/9	0.8909	$0{,}8882 = 1 - \dfrac{\sqrt{5}}{20}$
c − e	0.8000	4/5	0.7937	$0{,}7978 = 1 - \dfrac{\sqrt{5}+1}{16}$
c − f	0.7500	3/4	0.7491	0.7500
c − g	0.6666	2/3	0.6674	$0{,}6708 = 1 - \left(\dfrac{\sqrt{5}}{5} - \dfrac{\sqrt{5}-2}{2}\right)$
c − a	0.6000	3/5	0.5946	$0{,}5955 = 1 - \dfrac{\sqrt{5}+1}{8}$
c − b♭	0.5555	5/9	0.5612	$0{,}5928 = 1 - \dfrac{\sqrt{5}}{5}$
c − c	0.5000	1/2	0.5000	0.5

The "diatonic" intervals (fundamental note = length of string) start at the basis of the inscribed pentagon, from which they derive their relationships (Figure 8). They clearly extend to the meaningful intersection note a, with the seventh and eighth note falling into the upper, inscribed form. Instead of the note b, we see the original b flat.

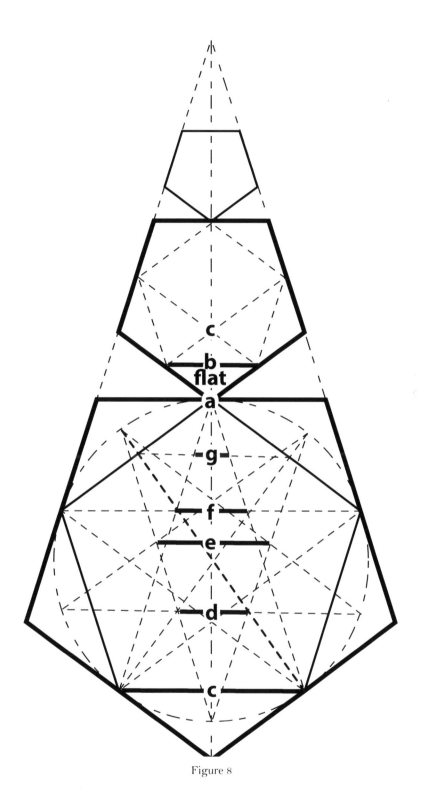

Figure 8

The section on Greek music in [Hugo] Riemann's *Musiklexikon* (*Encyclopaedia of Music*) [1889] draws special attention to the meaning of the note a in Antiquity: *"Through this boundary formed by the notes a' and A, as well as the central position of the note a, it becomes clear that the Dorian scale was heard as an A minor scale."* The pentagon further illustrates this statement.

If, as is the case, two tonal "genera" (the "enharmonic" and the "diatonic") are located very close to each other in the schema, then the third, the "chromatic genus" can be understood as an extension of the combination of the two former "genera".

As the *circle* with the decagon circumscribes the various tonal systems as a *unit*, there cannot be an abrupt contrast between the different systems. This clears away *a priori* all such objections that would otherwise be unavoidable. As does our modern system, Pythagorean music also harbours the pure notes, and the path from the Greek sounds to modern chordal language is not impossible. However, it is an arduous path, and only composers and artists who are enraptured by the exotic beauty of these lost harmonies will embark on it.

Only when these notes become tangible in new formulations do they begin to be generally convincing.

They will then communicate to our ears that which is connected, dreamlike, to the mighty symbols of Antiquity.

Bowed string instruments are probably the only option for the first musical experiments. Our current system of notation can be retained with the introduction of new signs. However, it is too early to go into such detail.

Through a restructuring, the ancient substratum can open up still unimagined perspectives. For now it is enough to elevate the insights of Pythagoras far above the suspicion of "unmusical speculation".

Editor's Note on Vituvius: The only complete book about architecture to survive from the ancient world is *De Architectura* (*On Architecture*) by Marcus Vitruvius Pollio, known always simply as "Vitruvius", who lived in the first century BC and was an architect and engineer. He was employed by the emperor Augustus in the rebuilding of Rome. Vitruvius was born between 80 and 70 BC and died sometime after 15 BC, aged between 55 and 65. We know little about him, but because he was fluent in Greek he was able to incorporate Greek as well as Roman architectural knowledge in his book, and in particular he could translate into Latin a passage from Aristotle's pupil Arixtoxenus about music, which is what interested Lucie Wolfer-Sulzer. There is no doubt that he was a genius, and his book is a bible to historians of architecture and archaeologists alike.

Sixteenth century editions of his work were profusely illustrated, and possibly the most fantastic such edition was that of 1567, published at Venice; some of its illustrations were foldouts and some even had additional pieces of paper to lift up and look underneath. As Eglantyne has available a copy of that edition in perfect condition, we may in the future issue a book containing its incredible illustrations, as they are a spectacular contribution to the history of art and architecture in themselves.

There are two readily available translations of Vitruvius's text into English: *Vitruvius, The Ten Books on Architecture*, translated by Morris Hicky Morgan, Dover Publications, New York, being a reprint of the original edition from Harvard University Press of 1914; it contains 68 rather basic illustrations. The other has Latin and English on facing pages, and is a two volume book in the Loeb Library series, with the translation by Frank Granger; the two volumes were published in 1931 and 1934, but are continually in print from Harvard University Press; they have a few basic illustrations at the back.

Chapter Four of the Fifth Book of Vitruvius begins like this (Granger's translation): "Harmony is an obscure and difficult branch of musical literature especially for persons unacquainted with Greek. If we wish to explain it we must use Greek words and some of these have no Latin renderings. Therefore I shall translate (as well as I can) from the works of Aristoxenus …" Among the things he reveals is that there are three kinds of scales: enharmonic, chromatic, and diatonic. He explains these in detail, and he then turns to the various chords and concords (concords being chords which are harmonious). He ends the chapter by saying: "And these concords are produced from the conjunction of sounds which in Greek are called *phthongi*." [*Phthongos* is the ancient Greek name for a note, or tone.]

Part IV contains much more discussion of Vitruvius and Aristoxenus, and further details of the works of those two ancient authors are to be found there.

Editor's Comment on Lucie Wolfer-Sulzer's innovative discoveries and new theory:

The above discoveries by the author are expressed in a compressed form and were sent in that form to colleagues who already had some education in music and musical theory, and who had some basic classical education as well. The majority of readers today have little or no classical education, as Latin and Greek are now taught in only a few schools. And as for musical theory, that has always been confined to a relatively small section of the public who have some reason to be involved with music. Therefore, the author's presentation of her findings will seem obscure to most contemporary readers, who will struggle with the background knowledge, not to mention the expertise involved with drawing and interpreting geometrical diagrams. But what the author succeeded in doing was to generate a system of musical tones by lines joining points in pentagonal diagrams, and pentagons are intimately connected with the Golden Section; a diagonal of a regular pentagon is in Golden Ratio to a side of the pentagon, and also when the diagonals of a regular pentagon intersect, they divide one another by the Golden Ratio. Thus, the author's tones might well be called 'Golden Section tones', and her system 'the Golden Section Tone System'. A key result of her system is to alter the frequencies of notes very slightly, and as she suggests, the results need to be explored further by playing with her tuning on a violin. Further discussion of the author's work in this area is to be found in the next text, Part IV.

PART III

PART IV

THE GREEK MUSICAL SYSTEM

Lucie Wolfer-Sulzer

Printed as a first draft manuscript
for private and preliminary circulation to colleagues

[Note: This was published just before the author's death in 1946.]

THE GREEK PITCH SPACE

The Roman scholar Vitruvius, who was a contemporary of the Emperor Augustus, included a chapter on Greek music at the very centre of his extensive work on architecture. It serves as an introduction to the subsequent sections on the construction of theatres, particularly in connection with acoustics.

Vitruvius based all his various deliberations strictly on classical art, even if he interpreted it in a rather superficial and idiosyncratic way. He clearly felt a stronger connection to architecture than to music and, as he himself admits, he followed the writings of the Greek philosopher and music theorist Aristoxenus in his reflections on the subject of music, which was foreign to him. Yet, on closer inspection it becomes clear that his disquisitions on this art form are more valuable than some of his chapters on architecture.

Aristoxenus of Tarentum (4th century BC) was a pupil of Aristotle. [*The Greek spelling of his name was Aristoxenos.*] He was closely associated with the intellectual heritage of Pythagoras, whose laws he staunchly defended in his writings[1]. Aristoxenus's (and, for that matter, Plato's) intolerance of any innovation and relaxation or detachment from the original system of the various art forms, which was strictly bound to ethics, is striking. Perhaps Plato and Aristoxenus clearly foresaw the decay of their culture in the "degenerate style".

For the purpose of the following investigations, we can conclude from these findings that Vitruvius's writings on music concern themselves with the *classical* Greek pitch space. *"Harmony* is a written description of the musical art, forbiddingly dark and unintelligible...". These are the opening words of the fourth chapter of the fifth book of Vitruvius (taken here from the excellent 16th-century translation of Vitruvius' works into German, entitled *"Vitruvius Teutsch"*, Nuremberg, 1548). The attributes "dark" and "unintelligible" indicate how unfamiliar the author was with the subject matter. Already during his lifetime, any connections with the intellectual foundations of Greek art had long been severed.

Vitruvius writes about the human voice, according to which the pitch space and the fixed pitches are determined. He mentions three original "tone species" (the *diatonic*, the *chromatic* and the *enharmonic*), and provides the Greek names of 18 pitches, which he separates into eight *stabiles* and ten *mobiles*. The pitches

---

1 *Editor*: Additional information on the works of Vitruvius was taken from a text by Dr. Fritz Wehrli, professor at the University of Zurich: *"Aristoxenus, Texte und Kommentar"* ("Aristoxenus, Text and Commentary"). Published by Benno Schwabe, Basel, 1945. It is Heft II of Wehrli's series of monographs, *Die Schule des Aristoteles, Texte und Kommentar*. As for Aristoxenus's enthusiasm for Pythagoreanism, 2019 saw the publication of Carl A. Huffmann's lengthy book *Aristoxenus of Tarentum, The Pythagorean Precepts*, Cambridge University Press.

in the latter category are said to achieve "consonance" by means of the various fixed intervals of the musical scales. For the organisation of the pitch space, the *tetrachord* plays a far more important role than the octave.

While this does hint at an outline of the system of the ancient pitch space, the details are thrown together in an undefined jumble. Of the three "tone species", only the *diatonic* – "*because it is a* natural *chant*" – has survived to this day. Even today, albeit as a truncated and tempered sequence of notes, it forms the basis of our music. The *chromatic* and *harmonic* can only be explained in part and thus have remained obscure.

If, as we assume, the *chroma* (the "coloured") is a mixture of "tone species", then we are essentially left with only two coexisting but separately developed "tone species": the *diatonic* and the *harmonic*. As late as the 16th century, the Italian [music theorist] [Gioseffo] Zarlino [1517-1590] touched on the forgotten tone series by claiming that in ancient times there existed both a "*harmonia propria*" and a "*harmonia non propria*"[2].

For the *harmonia*, therefore, we must assume pitches and note sequences that do not emerge from our familiar sequence of fifths and regular division of the string. In order to obtain any points of reference, we must first recall the Greeks' fundamental ethical principle. During the classical period, architecture, sculpture and painting were all subject to a unified geometric law, which was tied to the proportions of the *decagon* (double pentagon). In antiquity, this geometric figure, and its appertaining regular solid, the *dodecahedron,* was "*the image of the Gods, suffused by motion and life*" (Plato). [*Editor:* It was also claimed to be the shape of the Universe.] Well into the Hellenistic period, every important work of art was created according to this principle. It is therefore likely that the structure of the pitch space was also derived from it.

The study of the Greek temples provides an insight into the all-important metric proportions employed at the time. Examples of these proportions are the "golden section" (1 : 1.618), as well as $5 : \sqrt{5}$ (2.236), and 1 : 1056 (the limma in the Pythagorean system), etc.

Much as it was possible in antiquity, this hypothesis can be tested today using a simple taught string. When plucked, the whole string produces the fundamental note. (For the purpose of the following deliberations this will always be taken as *C.*) A shortening of the string leads to higher pitches. Thus, when halving

---

2 "*Rekenkundige Bespiegeling der Musiek*", Prof. Dr A. D. [Adriaan Daniël] Fokker, Gorinchem, 1944.
  [*Editor's note:* A French translation of Fokker's sequel to this work appeared in 1947: *Les Mathematiques et la Musique*, Martinus Nijhoff, The Hague, 1947 (a booklet of 31 pages), which contains much interesting information. He dos not refer to Zarlino again but he cites many other sources, some which are little known, and he takes account of the theories of Giuseppe Tartini, who, as is well known, opposed equal temperament.]

the string, one obtains the octave note *c*; 2/3 of the string provides the perfect fifth *g*; and 3/4 of the string the perfect fourth *f*. These consonant primary relationships must surely be assumed to form part of Vitruvius's eight *stabiles*, and consequently it is likely that they must be located not within a single octave, but within the entire – yet limited – pitch space. First, however, we will define the *mobiles* of the enharmonic mode that arise from the decagon.

By today's standards it is a rather peculiar procedure to measure out pitches on a given length, that is to say, a string, and to fix them visually. However, this is exactly how the Greeks proceeded (in addition to supposedly testing consonances with vessels filled to various degrees[3]). [*Editor's note*: The ancients had no way of measuring the frequencies of notes, so the two methods just mentioned were the only means available to them to get numerical data about tones; fortunately, the string lengths are the numerical inverse of the frequencies, so the ancients' numerical data was therefore reliable.]

For a string, the geometric figure that yields the metric proportions in the first instance is the pentagon, whose height is equal to half the length of the string and thereby corresponds to the pitch space of an octave. It is then possible, in line with the requirements of the Greeks, to *construct* the subsequent octaves in the ratio 1 : 2. However, for the purposes of subdividing the first octave, the pentagon must be thought of as being extended into a decagon (or as a double pentagon). From the lowermost point of the figure, which corresponds to one end of the string and the fundamental note C, the most important lines – the side of the decagon, the side of the pentagon, the diagonal of the decagon and the diagonal of the pentagon – emerge in regular angular distances (see the illustration on p. 152).

These lengths are subtracted from the length of the whole string, they are *cut off*; the Greek term *apotome* illustrates this process. The resultant pitches, relative to C, are a slightly flat c sharp, a slightly sharp e flat, a slightly flat g and a slightly sharp b flat. The c sharp corresponds to the diagonal of the pentagon, and, as the longest of the four segments, reaches into the region of the second octave. Therefore, it is shortened in the ratio 1 : 2 and thus returned to the first octave, where it is positioned next to the fundamental note as the *diesis* (the smallest interval). All four pitches that emerge from the primary proportions of the decagon are based on irrational numbers, which, as Plato describes in "Timaeus", can be extended into a long series by inserting the values of their mean proportions, i.e., their geometric means. However, this procedure yields too many possibilities, and consequently they must be ordered and restricted. How and where did the Greeks make their choices?

---

3 "Aristoxenus, Kommentar" ("Aristoxenus Commentary"), Fritz Wehrli.

Anyone who is mindful of the physical laws of sound will, from the outset, be more than sceptical about the notion of irrational divisions of the string. However, despite the insights won through science, the Greek theory must not only be judged from the viewpoint of the present. A number of mistakes have been made in the interpretation of the scant information passed down to us. All too often the octave was taken as the basis of analyses, and the fact that the Greeks defined the pitch space from the top to the bottom (as opposed to the other way round) was rarely taken into account. Few understood why the interval of the fourth was so important, and the pitches of Antiquity were simply dropped into our squashed, tempered tuning instead of being placed alongside the ancient, pure scale with its 19 pitches. This led to misunderstandings, to dissonances instead of harmonic sounds, and even to the denigration of Pythagoras and the system named after him. We must follow in the footsteps of the Greeks if we are to grasp their music. Plato is our great teacher, who not only bequeathed to us occasional statements about music, but everywhere in his writings alluded to the order of the universe and established a geometric law – which originates in unity – for all that has "become". He once said that one should admit that "the regular" (symmetry) is far more beautiful than "the irregular".

Given these postulates, it appears that the Greek pitch space is organised like a temple. The attempt to arrange it, as it were, architecturally has led to the following results. Its range is limited to two-and-a-half octaves, as it focuses on the best register for vocalists and the capabilities of the instruments of the day, which were very modest. It is divided into five tetrachords, each comprising four strings or pitches. The Greek terms for these tetrachords reveal the fact that the highest of them, the *hyperbolaeon*, provides the principal metric for the two following pairs, which in turn form an octave (see the table on p. 153).

Each tetrachord spans exactly a fourth, the *diatessaron*. Its outermost elements – the *stabiles*, the natural "immovables", which resound in harmony eight times over – support the entire pitch space. Within the inner octaves they coincide as *hypatē* and *nētē* while in the middle of the *diapason* (the octave) they are separated by an interval of a whole tone, for the upper tetrachord spans from c' to g, and the lower tetrachord spans from f to c. The pitch f coincides with the middle of the length of the octave and correspondingly is called *mesē* ("the middle one").

Thus an octave comprises two tetrachords *plus the whole tone between* the fourth and the fifth. Let us hear what Vitruvius has to say on the matter in his chapter on the acoustics of theatres:

"Nothing is to be placed in the middle, for the reason that there is no other note in *chromatico* (LWS note: also in *harmonia*) that forms a natural concord of sound."

The centre of the decagon, mysterious and untouchable, is located between the fourth and the fifth. In practice, this allows the omission of the pitches f sharp and g flat, which conveniently shortens the scale.

The large interval of the middle of the octave provides complete equilibrium to the two adjacent tetrachords. Much like in the construction of the temples of antiquity, their concordance is revealed: *each interval, even if it is of uneven or irrational placing, is echoed in the other intervals.* "Through distance the interplay of opposites is revealed." (Vitruvius).

Given this insight, one can progress towards selecting the (moveable or variable) *mobiles*. Regardless of whether we are dealing with enharmonic notes or those found in our ancient, pure tuning system, we can retain our familiar terms (and note system) for both scales. One may wonder whether these two types of scales are not essentially one and the same, with the difference that in antiquity the scale was derived from the decagon, while in later epochs only the approach of the regular division of a string was still known and therefore used – which led to the existing divergence.

Therefore, if the stable outer elements of the lowest tetrachord are C and f, the intermediate steps will be d and e, which can be lowered to d flat and e flat or raised to d sharp and e sharp. In the smaller, upper tetrachord that spans from g to c, the intervening pitches a and b can thus be changed into a flat and b flat or a sharp and b sharp respectively. Each displacement of a pitch results in the same procedure in the other tetrachords.

Something that seemed almost impossible at the beginning of this investigation, namely the consonance of irrational pitches, is now becoming a distinct possibility. However, for the Greek scale no unbroken circle of fifths can be employed – the pitch space is simply too limited – but nevertheless the two tone rows of the diatonic and enharmonic "tone species" flow in and out of each other in both symbolic and musical alignment. The ratio between the fourth and fifth is fixed, even in the irrational series, where, for example, the string lengths that are derived from the side of the decagon and the diagonal of the decagon respectively are exactly in the ratio of $2 : 3$ (0.5528 : 0.8292), that is to say, the interval of the fifth.

What remains is to determine the intervals for the forgotten "tone species" according to the information provided by Vitruvius. However, the intervals he mentions do not yield a reliable metric. The general confusion in this regard might be rooted in the fact that a whole tone in the Greek octave is not composed of two or four, but three intervals. Together with its outer *stabiles*, a tetrachord comprises seven *mobiles*[4]. Consequently, it consists of eight dieses, two of which

---

4   *"Aristoxenus, Kommentar"* ("Aristoxenus Commentary"), Fritz Wehrli.

are equivalent to a semitone, and three of which are equivalent to a whole tone. Having defined these relationships it is straightforward to determine the scales that correspond to Vitruvius's schema.

In the *diatonic*, intervals arise that correspond to our scales, that is to say, whole tone, whole tone, semitone (3/8 3/8 3/8) from the bottom up, while in the *chromatic* and the *harmonic* they must be applied in reverse order, from the upper octaves downwards. In the *chromatic*, the intervals through the *diatessaron* (tetrachord) should consist of semitone, semitone, tone-and-a-half (2/8 2/8 4/8), which can be written as a scale in this way: c − b − a sharp − g, and f − e − d sharp − C. In both tetrachords the pitches are consonant as slightly narrowed Pythagorean fifths. As required, this "tone species" possesses diatonic and enharmonic *mobiles*.

Things are quite different with the *harmonic*, which, according to Aristoxenus, is the younger of the two "tone species". The intervals are predetermined as whole tone, whole tone, diesis, diesis (3/8 3/8 1/8 1/8). Strangely this amounts to five pitches to a tetrachord, or ten in the *diapason*, the octave. *Ten* is the sacred number of the Pythagoreans. In the upper tetrachord c − g, the first interval reaches the point on the string that is determined by the *diagonal of the decagon*; the second interval reaches the point that is determined by the golden section, which is equal to b flat and a flat; while in the lower tetrachord the point of the *side of the decagon*, e flat, is reached. Apart from the *stabiles*, the entire scale consists of *enharmonic* pitches: c − b flat − a flat − g sharp − g and f − e flat − d flat − c sharp − C. This characteristic architecture probably also is the reason for its sacral function. *Terpander*, the celebrated Greek musician [of Lesbos, first half of seventh century BC], is reported to have used an instrument strung with ten strings (instead of the usual seven), and to have demonstrated the dignity of the Dorian-enharmonic mode in his solemn hymns[5]. According to Vitruvius, the "tone species" *harmonia* is of *"grand and splendid authority"*.

Across the entire pitch space, there are eight harmonic *stabiles*, and, taken together, a total of 17 pitches form an octave. Does not an image of the Parthenon with its eight frontal and 17 lateral mighty columns form before our eyes?

And so Vitruvius kept the best key to the gates of the Greek sound world to himself. In his writings there is a conspicuous lack of details about the various scales that are mentioned elsewhere by Aristoxenus. However, it is only a matter of different terminology within the same system. The *Lydian* mode corresponds to the *diatonic*, the *Phrygian mode* belongs to the *chromatic*, and the *Dorian* mode is equivalent to the *harmonic*.

---

5 c − *c sharp* − *d flat* − *d* − *d sharp* − *e flat* − *e* − *e sharp* − f   and
   g − *g sharp* − *a flat* − *a* − *a sharp* − *b flat* − *b* − *b sharp* − c

PART IV

For as long as the classical arrangement of the Greek pitch space remained untouched, transposition in the modern sense was impossible, but one could instead employ each of the three "tone species" in three different ways within each tetrachord. Among these variants one finds the *Ionian* and the *Aeolian* mode and arguably the *Syntonolydian* scale, while the mixed scales already violate the classical laws.

This interpretation of the Greek pitch space may well require historical and theoretical supplementation; a recreation of forgotten possibilities will involve much painstaking work. Nevertheless, the most beautiful sounds of antiquity have been made audible once again. Those who recognise their eternal value will discover new paths on which to ascend to the *Acropolis*.

Winterthur, January 1946

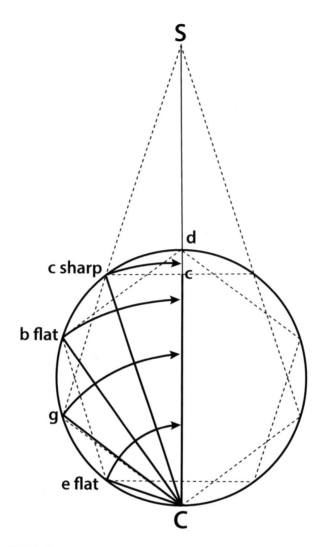

THE DIVISION OF THE STRING

C – S	=	whole string
C – c	=	half string
C – e flat	=	side of the decagon
C – g	=	side of the pentagon
C – b flat	=	diagonal of the decagon
C – c sharp	=	diagonal of the pentagon
C – d	=	height of the decagon

PART IV

THE GREEK PITCH SPACE

Arrangement through the tetrachord
(Details adjusted according to Vitruvius)

g	proslambanomenos		
———			
f	nete	hyperbolaeon	
e	paranete	"	
d	trite	"	
— c {	hypate	"	
	nete	diezeugmenon	
b	paranete	"	
a	trite	"	
g	paramese	"	
f	mese	meson	
e	lichanos	"	
d	parhypate	"	
	hypate	"	
— c {	nete	synemmenon	
b	paranete	"	
a	trite	"	
g	paramese	"	
f	mese	hypaton	
e	lichanos	"	
d	parhypate	"	
———			
C	hypate	"	

APPENDIX

Editor's further notes about Aristoxenus

Aristoxenus, originally from Tarentum in Sicily, was one of the brilliant young men whom Aristotle gathered as part of his research institute known as the Lyceum in Athens. Aristoxenus was outrageously prolific, reputedly having written 380 books. All but one, a portion of another, and some scattered fragments, have been lost. The surviving portions of one book, *Elements of Rhythm*, were only published for the first time in English translation in 1990. (Aristoxenus, *Elementa Rhythmica, the Fragment of Book II and the Additional Evidence for Aristoxenian Rhythmic Theory*, edited with introduction translation and commentary by Lionel Pearson, Clarendon Press, Oxford, 1990.) The only complete work of his to survive is his *Elements of Harmony*, also known simply as his *Harmonics*, published in English in 1902 and reprinted many times by reprint companies. (*The Harmonics of Aristoxenus*, edited and translated, with notes, introduction and index of words, by Henry Stewart Macran, Clarendon Press, Oxford, 1902.) It is not certain whether this work was a single work or was two originally separate works joined together, but it makes little difference. In 2019 a very ambitious book appeared, attempting to reconstruct Aristoxenus's lost work on Pythagoreanism by gathering and translating its surviving fragments and interspersing them with extremely lengthy commentaries. This mammoth book has 636 pages! (*Aristoxenus of Tarentum: the Pythagorean Precepts (How To Live a Pythagorean Life), An Edition of and Commentary on the Fragments with an Introduction*, by Carl A. Huffman, Cambridge University Press, 2019.) Anyone interested in knowing more about Aristoxenus can lose himself in that vast tome to great profit. Carl Huffmann was already notable as the leading expert on Aristoxenus, as editor of the seminal volume *Aristoxenus of Tarentum: Discussion*, Rutgers University Studies in Classical Humanities, Volume VII, Transaction Publishers, New Brunswick, New Jersey, USA, 2012. The Rutgers series, initiated by Professor William Fortenbaugh, consisted of volumes about the members of "The School of Aristotle". As mentioned in Part III of this book, the Swiss classical scholar Fritz Wehrli published the surviving fragments of Aristoxenus with commentary (but he did not translate them from the Greek and Latin): Fritz Wehrli. *Die Schule des Aristoteles, Texte und Kommentar* (*The School of Aristotle, Texts and Commentaries*), Heft II, *Aristoxenus*, Benno Schwabe & Co. Verlag, Basel, Switzerland, 1945. The first nine fragments deal with the life of Aristoxenus, and I have translated most of Wehrli's Commentary on that subject, though it is meant for classical scholars and is heavily academic in nature, so that general readers may find it perplexing:

"The admission of the Pythagoreans to the Freedom of the City of Tarentum [*i.e., as citizens*] (Fragment 2) belongs to the time of the Pythagorean power, which long preceded Aristoxenus. He was born a Tarentine, as was also Spintharos (according to Plutarch, *On the Sign of Socrates*, 592F, xxiii). Behind the statement concerning the father [of Aristoxenus] (Fragment 1) Müller [Carl Müller in *Fragment Historicorum Graecorum*] conjectures, with justification (p. 269), there is an attempt to bring about an accommodation between two variants; it has carried through in the tradition of the father Spintharos (fragment 51; see Sextus Empiricus, *Against the Mathematicians*, VI, p. 357), who in Fragment 1 is at the same time called the teacher of Aristoxenus in a series of musicians without any mention of their relationship. Aristoxenus himself (Fragment 54A) refers to himself as son of Spintharos for a characterizing of Socrates and, more remotely, probably in the context of an anecdote about Archytas [who was also from Tarentum] (see Fragment 30 and Commentary). Therefore it cannot be maintained in the face of these personal and direct references that either could have been the father and that in actuality Mnesias (or Muaseas) had been the father of Aristoxenus (compare scholion to Aristophanes *The Clouds*, 967: *Damprokleous tou Midõnos huiou mathēton*).

"The dating of the floruit of Aristoxenus in Fragment 1 seems to point towards the commencement of the reign of Alexander (compare Fragment 2 of Dicearchus), historically pretty well correct, for as candidate for the succession to Aristotle, Aristoxenus must have attained full manhood at the time of Aristotle's death in 322 BC. If he were the table companion of Neleus (Fragment 62 [from Lucian]), then he attended the School under Theophrastus (see Fritz, R.E., Volume XVI, 2280), who in Fragment 6 speaks of him as an experienced practitioner. [*Editor's note*: Fragment 6, re-edited, is now available in translation on p. 581 of Volume II of the *Theophrastus Sources* series from Rutgers University, edited by William Fortenbaugh. This gives the extraordinary story of Aristoxenus giving skilful musical therapy to someone who had been driven mad.] The dating in Fragment 9 [*i.e., the dating by Olympiad*] rests on a confusion with the poet of the same name of Selinus. The sojourn in Corinth (Fragment 31) [*Editor's note*: Fragments 30 and 31 are available in translation in Iamblichus's *Life of Pythagoras, available in more than one edition.*] one could take as a terminus post quem, of the banishment of Dionysius (see Diodorus Siculus, XVI, 70) in 344 BC. If Aristoxenus came from Mantinea (Fragment 1), then the Athenian years were preceded by a time in the Peloponnese. Regarding an encounter with Dicearchus who was living there as well as the personal and philosophical connections with him (Fragment 3), compare the Dicearchus Fragments 1 – 4; the Aristoxenus Fragments 118-21; 125 [*118-120b are from Cicero; 121 is from Martianus Capella*]; the sojourn in Thebes (Fragment 6) [*where he worked the musical cure*] allows of no further chronological determination. On chronological grounds the famous ancient musician Lampros (Fragment 1) cannot be the teacher of Aristoxenus [as suggested in this notice by Suda, which is the source of Fragment 1], and on the other hand since a younger bearer of the name is not known from Erythrae, the relation of a disciple from the admission of Fragment 76 or a list of great musicians would be suggested (compare for example Fragment 69A).

"Xenophilus (Fragment 1) is identical with the person of that name [a Pythagorean, from the Thracian Chalcidice region] in Fragment 19 [In fact, though Wehrli doesn't mention it, see also Fragment 25, where Xenophilus is described as Aristoxenus's "intimate friend". Huffmann has a great deal to say about Xenophilus in his book mentioned above.] upon whom Aristoxenus relies as an authority for the Pythagorean mode of life (Fragment 29 , in addition Fragments 20A and 20B and Commentary). If on account of this reference he is called his teacher (like Spintharos?), then the *diadochē* ["succession") is here meant as philosophical."

Carl Huffmann, mentioned above, announced in 2012 his intention to produce a definitive and greatly expanded collection of the Fragments of Aristoxenus, together with translations. Contacted in 2024, he replied that as he had retired and had not been able to complete that task, he had handed it over for completion to Professor Stefan Schorn of the University of Leuven in Belgium, who is Editor-in-Chief of the project known as *Die Fragmente der Griechischen Historiker* [*The Fragments of Greek Historians*]. He says (2024) that it will take some time because it is so complicated, due to the necessity to include all the fragments relating to the theory of music. Apparently when this enormous task is completed, the book will appear as a volume in the Rutgers University Studies in Classical Humanities Series, edited by Professor Bill Fortenbaugh.

The first English translation (with the Greek on facing pages) of the *Harmonics* of Aristoxenus was published as long ago as 1902: Henry S. Macran, *The Harmonics of Aristoxenus edited with Translation Notes Introduction and Index of Words*, Clarendon Press, Oxford, 1902. Modern reprints of this work began appearing in 2009.

A more recent English translation was published in 1989; this was done by Professor Andrew Barker (1943-2021), who at the time of his death was certainly the world's leading expert on ancient Greek music. In Volume Two of his two-volume work, *Greek Musical Writings* (Cambridge University Press, 1989) he included a translation of the *Harmonics* of Aristoxenus, though without the Greek text. His translation was massively annotated and preceded by an informative Introduction.

The general reader will not be familiar with these scholarly texts and discussions, so I have given this list of publications for those who wish to pursue the matter further. It is hoped that musicologists, musicians, and scholars of the history of music will find Lucie Wolfer-Sulzer's discoveries of the connection between Greek musical theory and the pentagonal geometry which is at the basis of Greek architecture and sculpture intriguing and suggestive. To elucidate an unsuspected unity of these three aspects of ancient Greek culture was in the Editor's opinion a magnificent achievement by the author.

R.T.